Taiwan's Environmental Struggle

Taiwan experienced a highly successful economic transformation in the last 50 years that produced one of Asia's genuine 'miracles' of modern development, in terms of improvement in per capita income and overall quality of material well-being for its citizens. The process, though, involved rapid industrialization and urbanization, and high mass consumption, that inevitably resulted in rapid degradation of the island's fragile air, water, and land, and produced some of the worst environmental pollution to be found anywhere in Asia.

This book examines the causes of Taiwan's environmental predicament, engaging in Taiwan's unique geological, geographical, demographical, political, industrial, historical, and economic circumstances. In addition, Jack Williams and Ch'ang-yi David Chang assess the efforts of the government, NGOs and private citizens to create a 'green' environmentally sustainable island, with a high-tech economy based on the silicon chip, the backbone of Taiwan's highly successful IT industry. Finally the authors discuss what can be done to improve Taiwan's environmental future.

As the first commercially available book in English on Taiwan's environmental problems this is an invaluable read for students and scholars interested in environmental studies, sustainable development and the island of Taiwan.

Jack F. Williams is Professor Emeritus of Geography at Michigan State University, USA. **Ch'ang-yi David Chang** is Professor of Geography at National Taiwan University.

Routledge contemporary Asia series

1 **Taiwan and Post-Communist Europe**
Shopping for allies
Czeslaw Tubilewicz

2 **The Asia–Europe Meeting**
The theory and practice of interregionalism
Alfredo C. Robles, Jr

3 **Islamic Legitimacy in a Plural Asia**
Edited by Anthony Reid and Michael Gilsenan

4 **Asian–European Relations**
Building blocks for global governance?
Edited by Jürgen Rüland, Gunter Schubert, Günter Schucher and Cornelia Storz

5 **Taiwan's Environmental Struggle**
Toward a green silicon island
Jack F. Williams and Ch'ang-yi David Chang

Taiwan's Environmental Struggle

Toward a green silicon island

**Jack F. Williams and
Ch'ang-yi David Chang**

Routledge
Taylor & Francis Group
LONDON AND NEW YORK

First published 2008
by Routledge
2 Park Square, Milton Park, Abingdon, Oxon OX14 4RN

Simultaneously published in the USA and Canada
by Routledge
270 Madison Ave, New York, NY 10016

Routledge is an imprint of the Taylor & Francis Group, an informa business

Typeset in Times by Wearset Ltd, Boldon, Tyne and Wear
Printed and bound in Great Britain by TJI Digital, Padstow, Cornwall

British Library Cataloguing in Publication Data
A catalogue record for this book is available from the British Library

Library of Congress Cataloging in Publication Data
A catalog record for this book has been requested

ISBN10: 0-415-44723-2 (hbk)
ISBN10: 0-203-92892-X (ebk)

ISBN13: 978-0-415-44723-2 (hbk)
ISBN13: 978-0-203-92892-9 (ebk)

Contents

List of figures vi
List of tables vii
Preface and acknowledgements viii

1 A crowded island: Taiwan's environmental setting 1

2 The perils of development: Taiwan's environmental problems and their causes 30

3 Back from the brink: resolving Taiwan's environmental problems 74

4 Whose land is it? Land use issues 135

5 Reflections: toward a green silicon island 166

Notes 182
Bibliography 204
Index 211

Figures

1.1	Taiwan	2
1.2	Taiwan: topography	4
1.3	Taiwan: drainage	5
1.4	Taiwan: population density	6
1.5	Taiwan: precipitation	8
1.6	Taiwan's aborigines	11
1.7	Taiwan: dams and reservoirs	17
1.8	Taiwan: urban and transportation patterns, 1950	24
1.9	Taiwan's coastal development projects	27
3.1	Government structure for environmental protection	92
3.2	Nature conservation areas in Taiwan	123
4.1	Cross-island highways	139
4.2	Land use in the Upper Tachia Valley, *c.*1970	141
4.3	Central Cross-Island Highway and the 1999 earthquake	149
5.1	Sustainable development	171

Tables

1.1 Indicators of Taiwan's development 10
1.2 Taiwan's environmental burden in comparison with other selected countries, 2001 25
2.1 Changes in PSI levels in Taiwan, 1985–2005 35
2.2 Changes in air pollutants in Taiwan, 1989–2005 35
2.3 Unhealthy PSI levels by urban/industrial regions in Taiwan, 1994 vs. 2003 36
2.4 Environmental burden in Taiwan, 1981–2003 37
2.5 Levels of pollution in Taiwan's rivers, 1987–2005 39
2.6 Varieties of harmful waste, 1997–2004 51
2.7 Taiwan's fisheries production/population, 2004 67
3.1 Domestic sewage control in Taiwan 106
3.2 Domestic sewage control in Taipei metro area 107
3.3 Projected primary energy demand, 2010 and 2025 130
4.1 Population and agriculture in Taiwan's slopelands, 2005 146
4.2 Crop planted area in Taiwan's slopelands, 2004 147
5.1 World economic forum growth competitiveness index, 2004–5 180

Preface and acknowledgements

This study of Taiwan's environmental struggle gestated for quite some time. As geographers, we have been studying Taiwan's environment and development since the late 1960s, and have collaborated on several projects over the years and made numerous field trips together to many parts of Taiwan. But with the reform era that began in the early 1980s, and the subsequent establishment of the Environmental Protection Administration in 1987 and the National Council for Sustainable Development in 1997, and a whole system of environmental agencies and organizations, as well as the emergence of the environmental movement among non-government organizations (NGOs) and citizens, we felt the time was ripe for a book-length study of Taiwan's environmental conditions, and the efforts being made to improve the environment.

The book really got under way in a formal sense, however, in late 2001 with a senior scholar research grant from the Chiang Ching-kuo Foundation, for which we are very grateful. That grant provided seed money for field work in Taiwan in the fall of 2001, with the longer range goal of eventually producing a book that would provide a comprehensive review of Taiwan's environmental struggle, especially in terms of its geography and policy implications. That book is now done. We hope that readers will find our analysis of Taiwan's environmental struggle both interesting and informative.

Chang was born and raised in Taiwan, and over the course of some four decades of academic work has become deeply knowledgeable about Taiwan's environmental situation, and is one of the pioneers of the environmental movement in Taiwan. Williams first went to Taiwan in 1969 to do dissertation research on a Fulbright Fellowship, and it was during that stay of more than two years that he came to know and become friends with Chang. Williams has returned to Taiwan innumerable times since then, including periods of residence as a visiting professor at National Taiwan University. The two have collaborated many times over the years, in a very fruitful and satisfying partnership. Williams has greatly appreciated the continuing assistance of Chang and the wealth of experience and knowledge that he brings to this study.

We believe this is the most thorough and comprehensive examination of Taiwan's environment to be carried out (in English) since the now-famous *Taiwan 2000* study, originally sponsored by the Asia Foundation (see Bibliography) that came out in 1989 (and for which Chang was Chair of the Steering Committee). That was a quite different kind of study, though, geared more toward an internal audience in Taiwan, especially government officials, with limited circulation outside of Taiwan, and more technically focused than our present study. We have deliberately tried to write this volume in as readable and lay-person-friendly a style as possible. For that reason, we have also deviated from the usual social science citation system by using endnotes in which both informative notes and citations are intermixed. We believe the main text flows more smoothly for the reader without the frequent interruption of citation sources in parentheses. Those who wish to know the sources, or seek additional explanations and details, can find them in the endnotes. At the end of the volume is a formal, comprehensive bibliography that includes all books, journal articles, and other publications used in the study. Romanization of Chinese proper names is always a headache, but especially in present-day Taiwan. For details on how we handled this vexing matter, see Note 11 for Chapter 1.

This study is aimed at both those who are already quite familiar with Taiwan and those who know very little or nothing about the island. We assume that Taiwan experts can skim over the explanations of Taiwan's basic social, political, and economic characteristics, and focus on those parts of the book that are new to them. At the same time, we hope that this volume will prove of interest and benefit to government officials and others in Taiwan who must deal with environmental matters. While we are sometimes quite critical, we also are fully cognizant of the extraordinary burdens that environmental officials must work under in Taiwan.

Thus, expressions of thanks or gratitude must first go to the (Taiwan) Environmental Protection Administration (TEPA) and its officials for their help. This is especially true for Chang, who has been closely involved with the TEPA since its establishment, as well as with numerous other government agencies over the years, and partnered with them in so many ways. Some of the brightest and most dedicated people in Taiwan work in the environmental realm, especially in the TEPA, and it has been our honour to know and work with them. We have also been helped by officials in many other branches of the government, such as the Council of Agriculture, the National Park Service, and others. These officials are far too numerous to cite individually here. The people of Taiwan are fortunate to have many dedicated and highly educated civil servants.

Our gratitude should also be extended to the many scholars, activists, and others working in the environmental arena outside of the government civil service, particularly in the myriad of NGOs, who offered their views and opinions formally in interviews or informally through conversations and interactions with the authors. These many individuals have truly been the vanguard of the environmental movement in Taiwan, and the people of Taiwan owe all of them a great debt for trying to stem the tide of pollution and environmental degradation.

We also want to thank Professor Bor-wen Tsai, in the Department of Geography at National Taiwan University (NTU), and his student, Ms Liu Chia-ling, who were responsible for producing the computer-generated maps and figures for this volume. They did a fine job, and their graphics help to make the text more understandable, especially the spatial dimensions of Taiwan's environmental struggle.

We should also thank other members of the Department of Geography at NTU, but especially the Chair from 2000–6, Professor Lin Jiun-chuan, for their generous support of our endeavours, and their gracious hospitality over the years.

We also thank Mr Wilson Ndovie, Computer Specialist in the Department of Geography at Michigan State University, for his technical assistance in preparation of the manuscript.

Finally, we take full responsibility for any mistakes in this study, especially misinterpretations or faulty analyses of facts and developments. Any academic study is filtered through the minds and experiences of the authors, and thus we inevitably bring our own biases and subjective viewpoints into the process at times, even while trying to maintain academic objectivity. We apologize in advance for any errors that did get past the screening process.

Jack F. Williams, Michigan State University
Ch'ang-yi David Chang, National Taiwan University
February 2008

1 A crowded island

Taiwan's environmental setting

When Chen Shui-bian was running for President in the late 1990s, he and his Democratic Progressive Party (DPP) chose the slogan 'Green Silicon Island' to symbolize what they stood for and hoped to achieve if they came to power in Taiwan. The 'green' represented not only purity of purpose and principles but also an environmentally restored and sustainable Taiwan. The 'silicon' stood for the silicon chip, the basic electronic device on which Taiwan's highly successful information technology economy was based. The DPP was claiming, in effect, that it would do all in its power to maintain Taiwan's remarkable economic progress within an environmentally sustainable system. It was a noble goal to aim for and it is likely that it won over some voters. Have they succeeded? Not by a long shot, at least not yet. Creating a Green Silicon Island will take much longer than a mere decade, and require the talents and commitment of many people beyond just the DPP. Taiwan has major hurdles to overcome to reach that seemingly utopian goal.

From an environmental perspective, Taiwan not only ranks as one of the more polluted places on earth, it also has a unique mix of factors that account for the island's relatively weak standing on the environmental front. Taiwan's environmental problems derive from several fundamental realities:[1] (1) The island is relatively small in proportion to its population (Figure 1.1). Hence, the impact of humans upon the environment is more intense than in less densely populated places. (2) The island experienced a highly successful economic transformation in the last half century that produced one of Asia's genuine 'miracles' of modern development, in terms of improvement in per capita income and overall quality of material well-being for its citizens. The process, though, involved rapid industrialization and urbanization, and high mass consumption, that inevitably resulted in rapid escalation in degradation of the island's fragile air, water, and land, and by the 1980s produced some of the worst environmental pollution to be found anywhere in Asia. (3) The island suffered for more than 40 years under an authoritarian one-party government that ran the island virtually unchecked in terms of development policies and allocation of the island's resources. The result was that abuses to the environment were buried under the rhetoric of higher priorities in the government's ongoing struggle with the Chinese government in Beijing. In short, the environment had to wait. Now, in

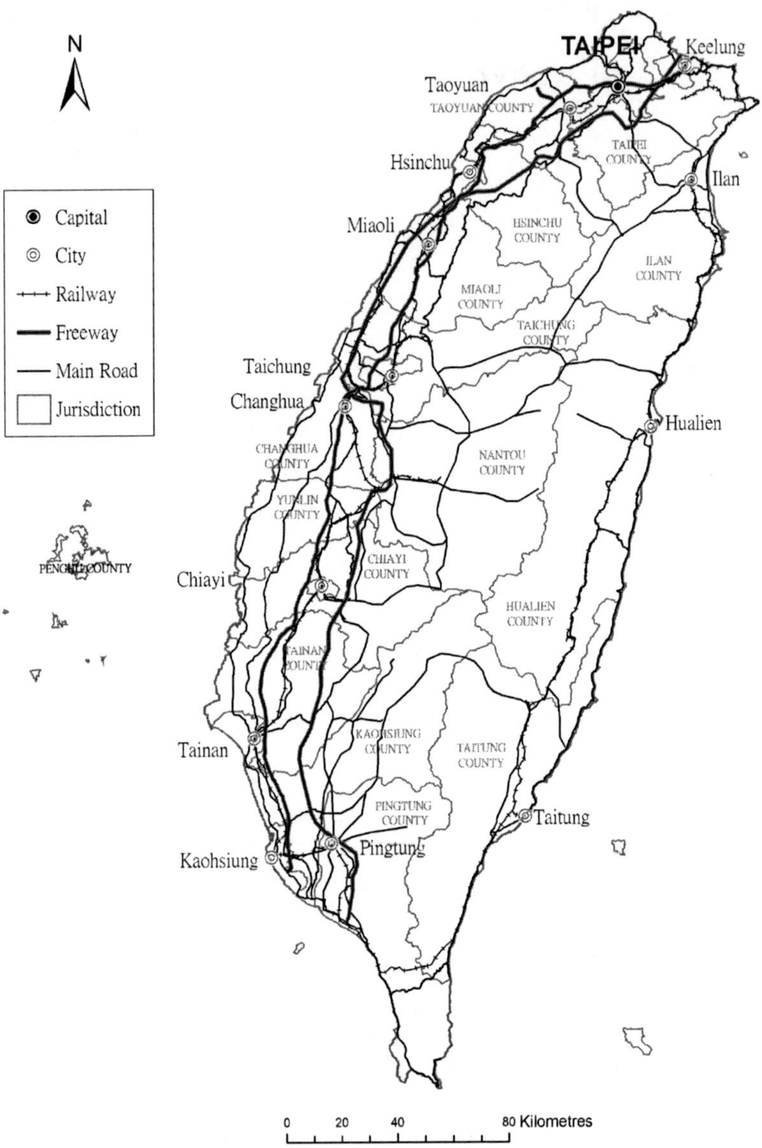

Figure 1.1 Taiwan (source: Institute of Transportation, 2006).

the last 10–20 years, under a new political system, Taiwan has been scrambling to catch up, to make amends and redress past abuses to the environment. The struggle has not been easy.

Taiwan, of course, is not alone in the world in having had these distinctive characteristics and experiences in the last half of the twentieth century. Many states are small, with high population densities.[2] Many have experienced rapid economic growth, while simultaneously seeing their environmental quality decline. Many have undergone periods of repressive political systems. South Korea is the one Asian state that perhaps has most closely paralleled the experience of Taiwan. Yet, there have been important differences between the story of Taiwan and that of South Korea (and other states). It would not be hyperbole to say that Taiwan is truly unique in modern Asia in the ways in which geography, economy, and politics have interwoven and interacted to produce the present environmental status of the island. What follows is the story of Taiwan and its environmental struggle.

Role of geography[3]

Taiwan's enormous geopolitical and economic importance in contemporary East Asia has a tendency sometimes to make one imagine that the island is much larger than in actuality. Taiwan has a total area of just under 36,000 sq. km, and is shaped somewhat like a tobacco leaf, running about 144 km at the widest point and 394 km in length (Figure 1.2). This makes it about half the size of Ireland, or nearly the size of the Netherlands or Switzerland.[4] On this island live a total of approximately 23 million people. Even if all of these people were evenly distributed over the island, Taiwan still would have one of the world's highest population densities. However, the distribution actually is highly skewed.

Taiwan is a geologically youthful, very mountainous island, dominated by a massive tilted fault block running north-east to south-west along the entire length of the island. More than 200 peaks rise to elevations over 3,000 metres. The highest peak, Yu Shan (literally 'Jade Mountain', a.k.a. Mount Morrison), reaches just under 4,000 metres (higher even than any mountain in Japan, including Mount Fuji). The slopes of this spine are steepest on the east, where mountains rise directly out of the sea in some places, resulting in limited level lowlands for human habitation. Thus, still today there are less than one million people living along the entire eastern side of the island. Erosion over geologic time has produced short, steep, shallow rivers flowing out of the mountain centre. Those flowing to the west produced a relatively wider coastal plain facing the Taiwan Strait and China along the entire western side of the island (Figure 1.3).

The bulk of the 23 million people (and most of the modern economy) reside there. In all, lowlands are in short supply. Alluvial plains below 100 metres occupy less than a third of the island. Hence, when referring to population density, it is more accurate to observe that Taiwan's people are actually occupying a land area slightly smaller than the US state of Delaware, or about half the size of Massachusetts. For the total land area, Taiwan's population density is already 622 persons per km^2, second highest in the world after Bangladesh, but greater than that of the Netherlands, with which the island is often compared, and more comparable to that of Japan, where lack of lowlands and unusual

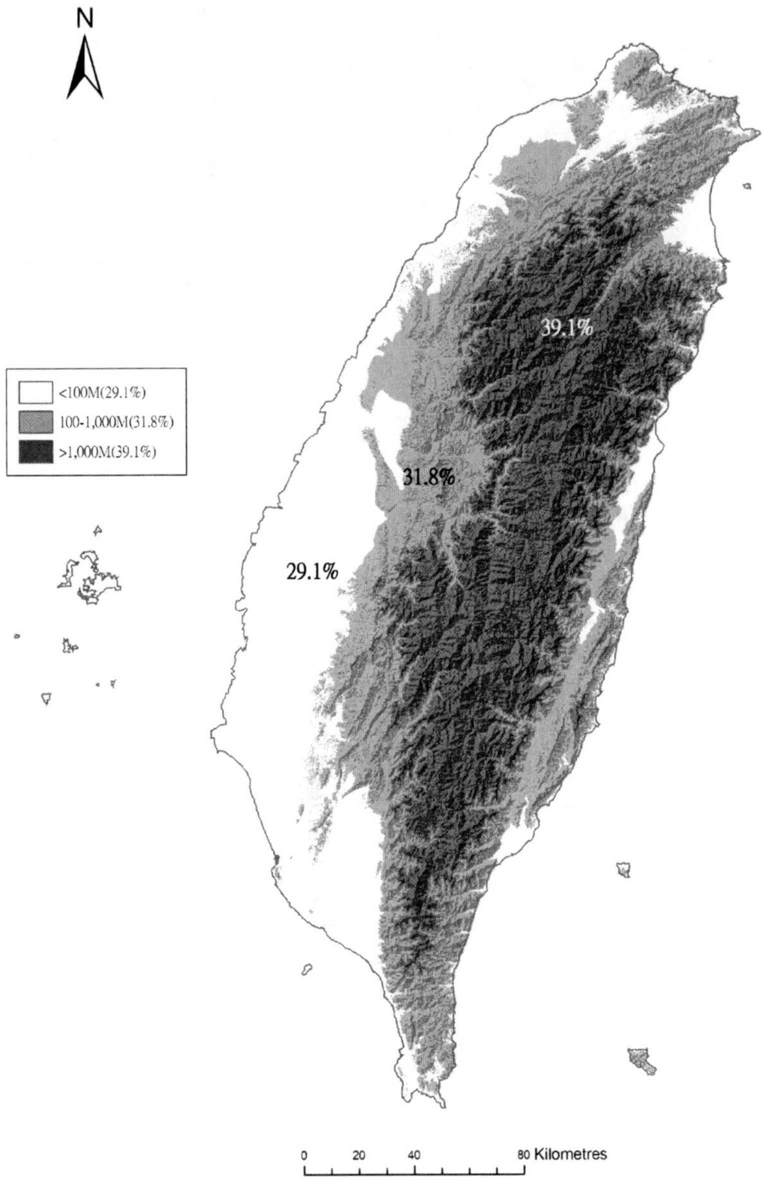

Figure 1.2 Taiwan: topography (source: *TEPA Environmental Yearbook*, 2004, p. 77).

urban distribution patterns also produce among the world's highest population densities. Taiwan's real population density, however, based on the actual land area occupied by most of the people, is closer to 2,500 persons per km^2. Further, in the major metropolitan areas of Taipei and Kaohsiung, the density soars to nearly 10,000 per km^2 (Figure 1.4). Under such high population

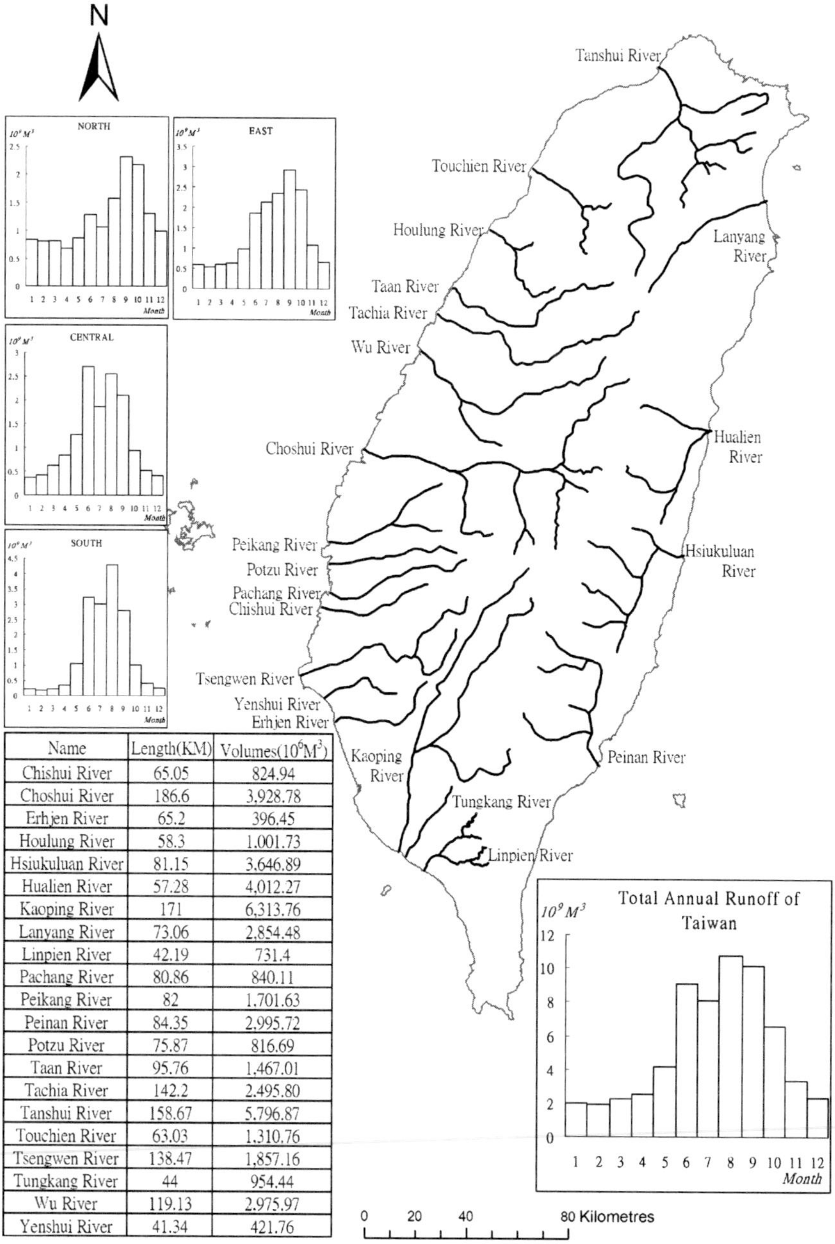

Name	Length(KM)	Volumes(10^6M^3)
Chishui River	65.05	824.94
Choshui River	186.6	3,928.78
Erhjen River	65.2	396.45
Houlung River	58.3	1,001.73
Hsiukuluan River	81.15	3,646.89
Hualien River	57.28	4,012.27
Kaoping River	171	6,313.76
Lanyang River	73.06	2,854.48
Linpien River	42.19	731.4
Pachang River	80.86	840.11
Peikang River	82	1,701.63
Peinan River	84.35	2,995.72
Potzu River	75.87	816.69
Taan River	95.76	1,467.01
Tachia River	142.2	2,495.80
Tanshui River	158.67	5,796.87
Touchien River	63.03	1,310.76
Tsengwen River	138.47	1,857.16
Tungkang River	44	954.44
Wu River	119.13	2,975.97
Yenshui River	41.34	421.76

Figure 1.3 Taiwan: drainage (source: *TEPA Environmental Yearbook*, 2004, p. 81).

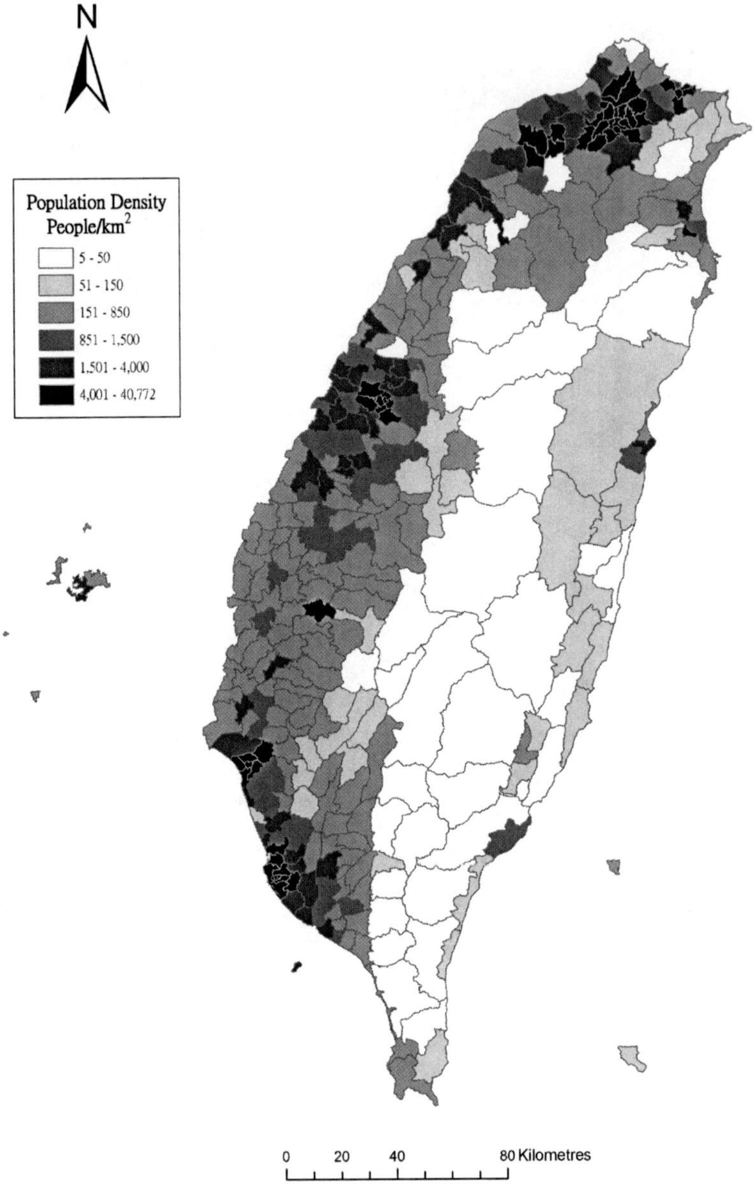

Figure 1.4 Taiwan: population density (source: Directorate-General of Budget, Accounting and Statistics, 2004).

density, it should be no surprise to find Taiwan has serious environmental imbalances.

Those imbalances are made worse by other aspects of Taiwan's geographical setting. The island straddles the Tropic of Cancer, or northern margin of the subtropical belt in East/Southeast Asia, between roughly 21° and 25° N. Lat., about the same latitude as the island of Cuba. Lying just off the south-east coast of China, though, means that Taiwan gets some continental influence, especially during the winter monsoon, and hence the northern half of Taiwan is relatively cool to mild in winter. Still, there is no frost in the lowlands (but there is occasional snowfall in the highest mountain elevations in winter), and hence a year-round growing season. Agriculture thus is intensive, with the need to eke as high yields as possible out of limited farm fields. Rainfall is abundant because of the subtropical monsoonal climate, albeit with distinct variations in rainy periods over the area of the island, November to February generally being the driest period. Annual precipitation is highest in the central mountains and north-east, where up to 3,000 mm[5] are normal. The mean for the entire island is about 2,500 mm (Figure 1.5). Yet, water shortages have actually become a reality in recent decades, especially in times of periodic droughts, because of so many people, a highly urbanized, water-thirsty economy, plus rapid run-off of surface water. The building of dams and water reservoirs thus has been a priority of governments since the Japanese era, but demand for water seems to grow faster than supply. Unfortunately, Taiwan also lies in the typhoon zone of Southeast and East Asia, and periodically suffers direct hits by devastating typhoons that sweep up out of the south-west Pacific and rake the Philippines and areas to the north, especially in late summer/early autumn. These systems can bring destructive winds and immense rainfall that produce serious erosion and flooding problems, as well as loss of property and lives. In short, water is a precious commodity with sometimes too much and sometimes too little, and hence there is a real need to use it wisely.

Lush natural vegetation, and some of the most beautiful virgin forest cover left in Asia, have felt the pressure of high population density and economic demands for forest products and use of marginal slopelands for agriculture and tourism. Not many people live in the uplands and mountains of Taiwan, but there is no mistaking the impact of humans on that major realm of the island. Compounding all of this is the island's location on the circum-Pacific 'Ring of Fire', or tectonically active zone that follows the junctions of the continental plates all around the Pacific Rim. The result is frequent earthquakes whose impact is magnified by high population density, steep slopes, and often improper use of slopelands. The September 1999 quake in central Taiwan was the most recent devastating earthquake to hit Taiwan, causing massive destruction and heavy loss of life and injuries.[6]

Earthquakes, typhoons, and floods are natural calamities that humans have no power to prevent, of course. But in a high population density island such as Taiwan, the government and people have an obligation to do all that is humanly possible to design the layout of cities, transportation networks, and economic

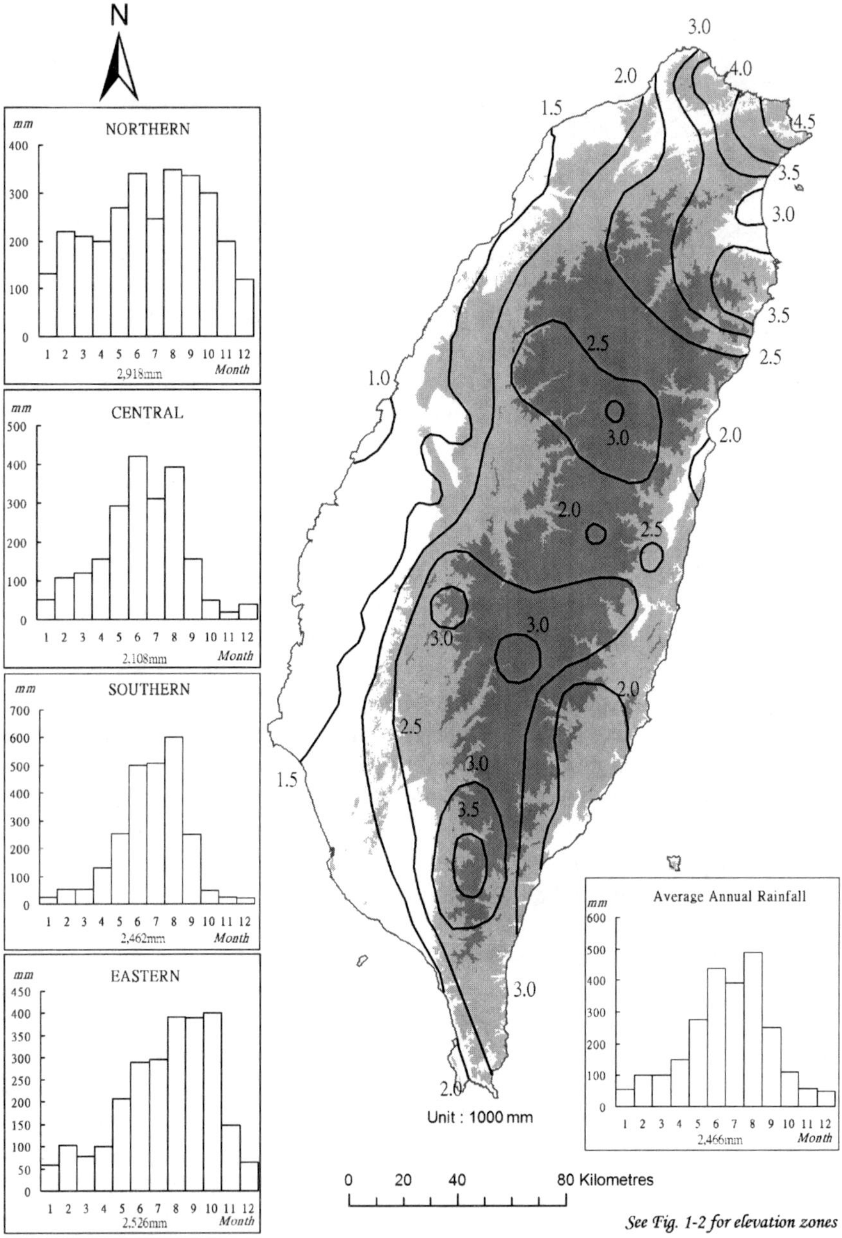

Figure 1.5 Taiwan: precipitation (source: *TEPA Environmental Yearbook*, 2004, p. 80).

systems to try to minimize the impact of such calamities when they occur. Unfortunately, this has not always been the case in Taiwan, and there is much that remains undone in terms of protecting Taiwan and its people from the devastation of natural calamities.

Role of the economy

Taiwan's dramatic transformation from a primarily agricultural colony of Japan in 1945 to one of the most powerful industrial economies of Asia by the end of the twentieth century continues to amaze observers and has spawned a huge literature by economists and others trying to explain it.[7] This is not the place to go into a lengthy examination of this complex process. But it is important, in trying to understand Taiwan's environmental situation today, to note the main features of the island's economic transformation and how that affected the environment.

Some basic statistics provide solid evidence of the transformation. In 1949 when the ROC government completed the retreat to Taiwan, after losing the civil war with the Communists on the mainland, Taiwan was wracked with social and economic turmoil, seemingly destined soon to fall to a communist invasion and incorporation into the newly established Peoples Republic of China. No one could have predicted what would happen in the next few decades. As a result of complex geopolitical events, Taiwan remained independent of China and proceeded to develop politically and economically in its own unique way. By 2004, as Table 1.1 reveals, Taiwan had been transformed in many ways.

Unlike the PRC, which established a centrally planned command economy similar to that of the Soviet Union, Taiwan opted for a mixed economy in which the central government played a guiding role (especially in the early period), but the focus was primarily upon establishing an optimum environment for private enterprise.[8] The economic system reflected and was in part a direct consequence of Taiwan's human resources, operating within a unique political environment.

Taiwan's human resource

Vital to that economic system has been the island's population, composed of four main groups. Although the vast majority (about 98 per cent) of the people are of the same Han ethnic background as the bulk of the mainland's population, Taiwan's Han are divided into three distinct groups. The largest are the Hoklo,[9] about 75 per cent of the total population, who are descendants of immigrants from southern Fujian on the mainland. Their native dialect is *Min-Nan Hua* (Southern Min language), the same as that of southern Fujian today, but known on Taiwan as the 'Taiwanese' dialect. The second group of people is the Hakka, about 12 per cent of the population. Originally from Fujian and Guangdong, they have their own dialect, but most also learned to speak Taiwanese in order to survive on Taiwan. Collectively, these two groups, Hoklo (or Southern

Table 1.1 Indicators of Taiwan's development

	1952	*2003/4*
Population (millions)	8.128	22.605
Population density (persons/km^2)	235	625
GNP (US$millions)	571	316,700
PCI (US$)	196	13,529
GDP by industry (%)		
Agriculture	32.2	1.74
Manufacturing	12.9	29.54
Services	48.1	68.72
Employment by industry (%)		
Agriculture	56.1	7.3
Manufacturing	16.9	34.8
Services	27.0	57.9
Exports (US$millions)	116	174,010
Imports (US$millions)	187	167,890
Agriculture's share of exports (%)	91.9	1.5
Manufacturing's share of exports (%)	8.1	98.5
Urban population (%)	20.7	77.7

Source: *Taiwan Statistical Data Book* (various years); Directorate General of Budget (2004).

Fujianese) and Hakka, are known as the 'native' Taiwanese, essentially a political term that refers to those Han immigrants who came to Taiwan before 1945, and their descendants. These two groups started immigrating to Taiwan many centuries ago, but most of them came from the 1600s onward. The third group of Han is the 'Mainlanders', also a political term, referring to immigrants who arrived on Taiwan primarily during that narrow window of time between 1945 and 1949, when the Republic of China (ROC) transferred operations to the island.[10] The Mainlanders are a very mixed group from many parts of China and constitute about 10 per cent of Taiwan's population today. In addition to their own local dialects brought with them to Taiwan, they also mostly speak Mandarin Chinese, the national language.[11] The fourth group in the population is the aborigines, or indigenous peoples, a mere 2 per cent of the population, but consisting officially now of 13 tribal groups that are still distinguishable and not yet totally assimilated.[12] At less than half a million in total number today, the aborigines are of complex origin, having come originally from southern China and Austronesia, and possibly other places, as early as 12,000–15,000 years ago, and are the true original inhabitants of Taiwan (Figure 1.6). In general, those from southern China tended to settle in northern and central Taiwan, while those from Austronesia concentrated more in southern Taiwan. In modern times, the aborigines have played the least important role in Taiwan's economy and politics. About half of them live in the mountainous interior and eastern part of Taiwan. Most of the aborigines are heavily assimilated today, but there is a growing assertiveness of late for aboriginal identity and rights.

In terms of demography, Taiwan's population of 23 million is now growing

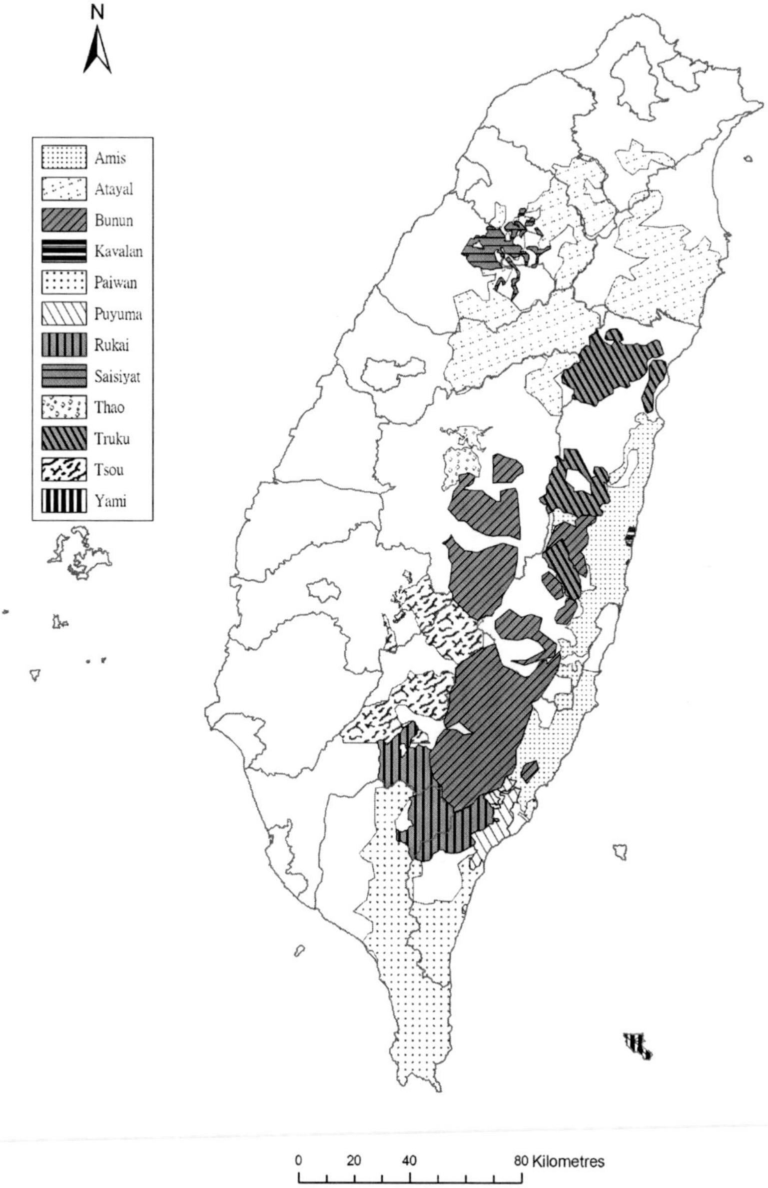

Figure 1.6 Taiwan's aborigines (source: Academia Sinica).

very slowly, having moved very close to the end of stage three of the demographic transition, in which births and deaths are nearly equal. The population growth rate is currently around 0.4–0.5 per cent; hence, the average age is rising, and the population is maturing, as the number of elderly people increases as a proportion of total population. Taiwan thus is sort of midway between fully mature countries, such as Britain, France, Japan, and Germany, and more demographically youthful countries, such as China, South Korea, Thailand, and the Philippines. As with a number of maturing countries around the world, Taiwan is beginning to be concerned about the social and economic consequences of a stabilizing, and eventually declining, population base, without significant immigration. Limited numbers of overseas Taiwanese have returned to Taiwan in recent years, with the establishment of democracy, and there is some illegal immigration, mostly mainland Chinese who overstay tourist visas and try to blend in. But these numbers are not enough to offset declining natural population growth.[13]

Opinions are deeply divided in Taiwan over this demographic trend, with scientists and environmentalists arguing that it may be a good thing Taiwan faces a declining population, because it could ease pressure on the environment and resource base. Others argue that the economy is geared to the paradigm of a growing population and growing consumer demand. This is a critical international issue affecting an increasing number of countries, with important ramifications for many aspects of national development, not least the problems of environmental improvement and protection. For one thing, ageing, stabilizing (or declining) populations use resources and view the environment in different ways from youthful, expanding populations.[14] Moreover, the government and society must readjust their focus on providing sufficient health care and other services needed by an expanding elderly population. Taiwan's elderly population currently is about 10 per cent of the total and is projected to reach 13 per cent by 2016, when it will start to outnumber the proportion under the age of 15.[15] The government more and more seems to be leaning toward seeing a declining population as a real problem and is seriously thinking of providing child-raising subsidies, as well as other measures, to boost population growth.[16]

In any event, with its unusual, politically defined human resource mix, Taiwan has undergone a series of transformations in its relatively short history (i.e. 'short' in comparison to the lengthy history of China). The changes since 1949 are best understood within the context of earlier foundations on which modern Taiwan has been built, and the roles that the various population groups have played in these successive transformations.[17]

Taiwan's first technological and social transformation

Although sporadic Chinese immigration to Taiwan had occurred for many centuries, it was really the Dutch who put Taiwan on the map, so to speak, during their brief period of colonial rule (1624–62). Until the Dutch arrived, Taiwan was controlled by no one and had a small indigenous population and a handful of Hoklo settlers, in a primitive, largely frontier wilderness. Even during the Dutch period, their control was limited to a tiny portion of the island. The Dutch were among the first Western explorers and colonizers in the region, and saw in Taiwan (then known also as 'Formosa', after the Portuguese name for the island) an opportunity to establish an agricultural trading enclave in Asia. They established a toehold on the south-west coast, near the present site of Tainan. The Dutch needed Chinese farmers to work for them in establishing rice and sugar cane production, and hence encouraged Chinese immigration from Fujian across the Taiwan Strait. Thus began organized Han Chinese immigration to the island, and the first technological and social transformation, a modest start to be sure, but nonetheless the beginning of Taiwan's economic development. Although the Dutch impact was not only brief, but also limited in geographical scope, the Dutch paved the way for others to follow.

Dutch rule might have lasted much longer but for the collapse of the Ming Dynasty on the mainland in the early 1600s. Ming loyalist forces, under Cheng Ch'eng-kung, fled to Taiwan with the hope of using the island as a sanctuary to rebuild strength and eventually return to the mainland to evict the alien Manchus, who had just conquered China and established the Qing Dynasty (1644–1911). The Dutch on Taiwan were no match for the Cheng forces and thus their short tenure came to an abrupt end, with little lasting impact except for having shown the agricultural potential of Taiwan and the settlement possibilities for further Chinese immigration. The Chinese population of the island has been estimated at somewhere between 40,000–100,000 in 1683, with perhaps another 100,000 aborigines. Most Chinese were still concentrated in the south-west, but beginning to probe northward into other parts of the coastal plain. Agriculture continued to expand. However, Cheng rule (1662–83) did not survive long either. Political infighting on Taiwan among the Cheng forces and the growing strength of the Manchus resulted finally in the Qing forces successfully invading Taiwan in 1683, and formally incorporating Taiwan into the Chinese empire for the first time.

Hence, China's rule of Taiwan is formally dated from 1683, one of the last pieces of territory to be added to the empire. The Qing Era on Taiwan lasted just over 200 years (1683–1895).[18] Initially, the Qing attempted to restrict further emigration to Taiwan, which was a prefecture of Fujian for most of this era, to little effect. Hakka followed the Hoklo to the island, as they spread up the entire west coast plain. The Hoklo pushed out the aborigines to occupy the best land; unassimilated aborigines moved to the mountain interior. Hakka occupied the inferior agricultural land, laying the foundation for rivalry and tensions with the dominant Hoklo, with distinctly different settlement locations between Hoklo and Hakka. The east coast valley and mountain interior were still untamed wilderness controlled by headhunting aborigines. By 1811 Taiwan's population reached two million, and another million were added by the end of this era in 1895. Most Taiwanese lived in rural communities, isolated from each other and with very limited contact with Chinese administration. Taiwan earned a reputation as a lawless, rebellious place, and not very healthy because of endemic tropical diseases, such as malaria. There were few urban settlements, mostly modest towns, such as Taipei in the north.

From an environmental viewpoint, the Qing era was significant in that agricultural settlement proceeded throughout the west coast plain, and gradually pushed into the foothills of the mountainous interior. As settlement spread through the lowlands, the natural forest cover steadily disappeared, replaced by rice paddies and farm fields, farmsteads, and rural villages. The remaining natural forest was primarily found in the interior slopelands. Both aborigines and Chinese settlers made increasing use of these slopelands, especially along the frontier boundary between west coast plain and interior mountains. One of the tree species especially sought after was the camphor tree, which is native to Taiwan. Camphor became a prize commodity and export product of Taiwan in the latter half of the nineteenth century.[19] The Qing government tried to make the natural forest cover a government monopoly and to forbid illegal cutting of camphor trees, with limited success. This often led to bloody conflicts between Taiwanese and aborigines, as the Taiwanese pushed deeper and deeper into the mountains in search of camphor trees, initiating a trend for ever-increasing pressure on the fragile resources of the mountains, both flora and fauna, that would reach major proportions during the twentieth century.

The Qing era was notable for a relative lack of significant economic and social advancement on Taiwan. It was precisely that relative neglect by the Qing government that contributed to Taiwan being separated from the Chinese empire after just 200 years. It then fell to a second group of non-Chinese colonizers, the Japanese, who 'won' Taiwan as a prize in their first war with China in 1894, to push the transformation of Taiwan to new heights.

Taiwan's second technological and social transformation

The Japanese colonial era (1895–1945) was arguably the most influential period in Taiwan's pre-1949 history.[20] Japan was newly arrived as a member of the world's colonial powers. Taiwan was their first foreign colony (if one does not include incorporation of the Ryukyu Islands into the Japanese Empire in the 1870s). Japan was determined to make a success of their venture, partly to prove to the Western powers (and to other Asians) that Japan was fully equal to any Western nation. Their initial interest in Taiwan was to more fully develop its agricultural potential so that the island could help supply essential agricultural products, especially rice, sugar, tropical fruits, timber, and others, to the homeland as the empire expanded and Japan's population grew. Later, as Japan militarized and began preparations for what was to become the Pacific War in World War II, limited, small-scale industrialization was begun and the island was developed as a forward base for the Japanese military. The invasion of Southeast Asia in 1941 was launched from bases in Taiwan.

Japan was highly successful in its economic initiatives in Taiwan, and thus this half-century era can rightly be described as Taiwan's second technological and social transformation. Agricultural production expanded enormously, especially for rice and sugar cane, partly through increased yields brought about by more modern agricultural practices and greatly increased irrigation, partly through expansion of the cultivated area. The west coast plain became more or less fully utilized during this era, and the first significant settlement and development of the east coast valley and coastal plain also took place, as Japanese and Taiwanese settlers moved there. The urban pattern for the island as we know it today began to take shape, with the colonial capital established in Taipei (known then as *Taihoku* in Japanese), served by the north-east port of Keelung, and development of the major port city and naval base of Kaohsiung (*Takao* in Japanese) in the south-west. Other secondary cities and towns emerged or further developed, also, such as Taichung in the central west coast plain, and the key east coast ports of Ilan, Hualien, and Taitung. Connecting these various urban centres was a growing network of railways and roads laid out by the Japanese.[21] Tropical diseases were brought under control or completely eliminated. In short, the basic spatial pattern of Taiwan's economy and settlement took shape during the Japanese era. Taiwan's post-1949 development would likely not have been so successful so fast if it had not been for the solid foundation laid by the Japanese.

The people of Taiwan responded in various ways to Japanese rule. There was indeed a social transformation. Those who resisted were ruthlessly suppressed or left the island. Most Taiwanese, however, accepted Japanese rule and tried to make the best of the situation, focusing on family advancement and accommodation with Japanese demands. Many Taiwanese prospered, some by going to Japan for education and developing links with Japan. For most Taiwanese, material well-being improved during the Japanese era, at least until wartime shortages and Allied bombing in the 1940s finally had an impact. Even then, in 1945 the average standard of living in Taiwan was measurably higher than that on the mainland, one of the contributing factors to the sour relations that developed between Taiwanese and Mainlanders after 1945. Moreover, the Taiwanese had been heavily assimilated, or 'Japanized', accelerating especially in the 1930s, adopting the Japanese language, dress, and other customs. Japanese architecture, house style, and food were widely adopted. The legacy of that cultural conversion can still be seen in Taiwan today, and contributes to the lingering sense by many Taiwanese of separateness and being 'different' from the mainland. Economic and cultural links with Japan remain very strong to this day. Taiwan is a favourite destination of Japanese tourists, for example.

The aborigines suffered the most under Japanese rule. The Japanese established a fortified frontier line around the mountain interior where unassimilated aborigines were concentrated, and gradually tightened the noose over the decades. There were many bloody encounters. The Japanese were interested in exploiting the forest and water resources of the interior, of course, and the aborigines were simply obstacles in the way. For example, the Japanese built a dam in central Taiwan in 1934 that greatly expanded Taiwan's largest freshwater lake, Sun-Moon Lake, forcing evacuation of population living around the lake but establishing a major water reservoir for agricultural irrigation and a site for recreation and tourism that is still important. The Japanese thus started the process of dam building on Taiwan's steep rivers to tap the hydroelectric and irrigation potential that came to play a key role in the island's tremendous agricultural development in the twentieth century, especially in the 1920s and 1930s and again in the 1950s and 1960s. But from an environmental viewpoint, these water resource projects sometimes had serious repercussions, and in recent years have become centres of controversy, as activists have focused public attention on the environmental costs of these structures.[22] The Japanese also began the process of penetrating the mountain interior with roads and the first railway into the interior. This was a narrow-gauge line up 2,270-m high Mount Ali in the south, which allowed the Japanese to exploit the interior timber resources, and to develop the area as a prime mountain recreation site.[23] This process also was enormously expanded after 1949, with construction of cross-island highways and countless feeder roads for logging and slopeland cultivation projects, thus becoming another focus of controversy in use of mountain resources. In the lowlands, one of the most famous Japanese projects was development of the Chianan irrigation system in what is today Chiayi and Tainan counties. The Japanese built the system in the 1920s, drawing water from reservoirs and an elaborate system of canals that they constructed. The Chianan Plain thus became a key agricultural area of Taiwan, greatly boosting the island's production of rice and sugar cane (Figure 1.7).[24]

Environmentally, thus, the Japanese era was one of maximizing exploitation of the natural resources of Taiwan. While this was done primarily for the benefit of the Japanese colonial forces in Taiwan and for the home islands, the Japanese were no fools and in general did not recklessly exploit the island without consideration of the long-term impact of their policies. The Japanese thought they were going to be in Taiwan forever, and realized that they needed to protect the forest cover, soils, water, and land resources of the island if their model colony were to prosper over the long haul and serve its assigned purpose within the Empire. In their favour, of course, was the fact that the population density even at the end of the Japanese period was still quite low compared with today, and the population was still overwhelmingly rural. Chinese immigration to the island virtually ceased during the Japanese era, while only limited numbers of Japanese settlers voluntarily emigrated there (in spite of a vigorous campaign by authorities to recruit Japanese settlers in the home islands to relocate as pioneers to the colony in Taiwan). There were less than 200,000 Japanese on the island in 1945 and virtually all of them were repatriated back to Japan. The total population of the island was about six million then, compared to just over three million at the start of the Japanese era. Hence, it took roughly 50 years to double the population, almost entirely from just natural increase, not immigration.

Taiwan's third technological and social transformation

In 1945 the Japanese were defeated, the Empire collapsed, and Japan forced to return Taiwan to the ROC government, which was seeing the civil war with the

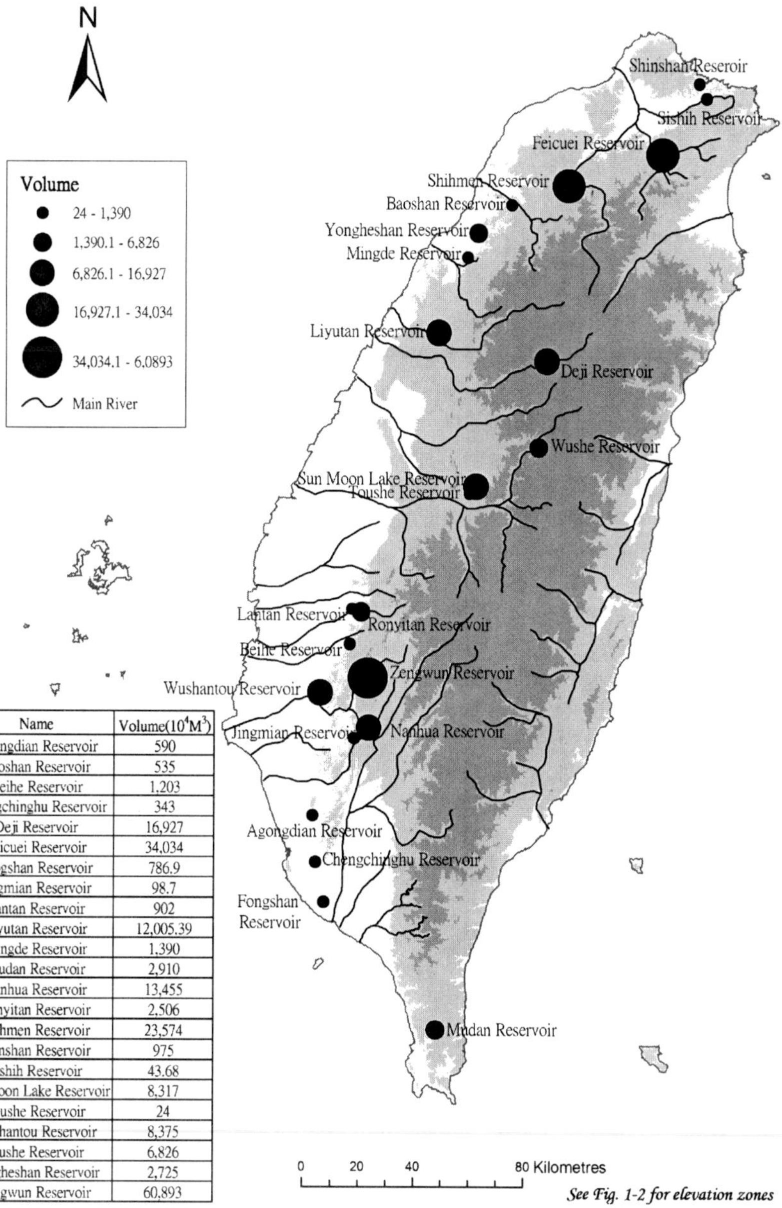

Name	Volume(10^4M^3)
Agongdian Reservoir	590
Baoshan Reservoir	535
Beihe Reservoir	1,203
Chengchinghu Reservoir	343
Deji Reservoir	16,927
Feicuei Reservoir	34,034
Feogshan Reservoir	786.9
Jingmian Reservoir	98.7
Lantan Reservoir	902
Liyutan Reservoir	12,005.39
Mingde Reservoir	1,390
Mudan Reservoir	2,910
Nanhua Reservoir	13,455
Renyitan Reservoir	2,506
Shihmen Reservoir	23,574
Shinshan Reservoir	975
Sishih Reservoir	43.68
Sun Moon Lake Reservoir	8,317
Toushe Reservoir	24
Wushantou Reservoir	8,375
Wushe Reservoir	6,826
Yongheshan Reservoir	2,725
Zengwun Reservoir	60,893

Figure 1.7 Taiwan: dams and reservoirs (source: Water Resources Agency, Ministry of Economic Affairs).

Communist insurgents resume. There was a brief resumption of Chinese immigration to the island starting in 1945, with the arrival of the 'Mainlanders', as noted earlier. Initially, the government moved a small number of troops and administrators to the island, but then as the Communist forces rapidly took over the mainland, the remainder of the ROC government, under Chiang Kai-shek, and a large military force, many with dependents, fled to Taiwan, a transfer completed in 1949–50. Altogether, upward of a million or more Mainlanders immigrated in this four-year period, boosting the island's population to about 7.5 million by 1950. Most of the Mainlanders settled in the Taipei area, occupying structures abandoned by the departed Japanese, or living in squatter settlements. Limited numbers of Mainlanders settled elsewhere on the island, mostly administrators (some with dependents) in charge of now ROC government businesses and concerns, such as the various plantations operated by the monopoly Taiwan Sugar Corporation.[25] A distinct population distribution – a Mainlander-dominated north vs. a Taiwanese-dominated south, based on ethnicity and politics, became established on Taiwan, a pattern that still influences many aspects of the island's development, including environmental issues.

Thus began the Republic of China (ROC) era, which continues to the present day. This era is divided into two parts: the first part was the Martial Law period (1945–87), and the second is the Post-Martial Law period (1987 to the present), these two halves defined by the internal politics of Taiwan. In the first part, Taiwan underwent its third technological and social transformation. In those years, immigration to Taiwan virtually ceased, as the mainland underwent the convulsions of establishing a Maoist society largely cut off from the outside world. Taiwan was now the rump state of the 'Republic of China', under the Kuomintang (KMT) or 'Nationalist' party rule, 'temporarily' in exile on the island, but dedicated to using Taiwan as a place to regroup and rebuild strength, for the eventual day when the ROC government would return to the mainland, throw out the Communist 'bandits', and resume administration of all of China.[26] 'Recovery of the Mainland' was a quixotic dream that would never be realized, but it fuelled the policies of the government that played a key role in further transforming Taiwan and turning it into a powerful and rich urban/industrial state.[27] Unfortunately, the environment often proved a victim of the steadfast policy to recover the mainland.

Each of the decades, from the 1940s on, came to symbolize a phase in the third technological and social transformation of Taiwan.

1940s

The late 1940s were a period primarily of consolidation of KMT control in Taiwan, but also the beginnings of economic reconstruction, and the start of land reform. The ROC government recognized the bedrock importance of bringing rural prosperity to Taiwan as the first step in not only transforming Taiwan's economy but also in winning the hearts and minds of the native Taiwanese, after the bitterness of the 28 February 1947 tragedy.[28]

1950s

The 1950s was a decade of intense focus on agriculture, with continuation of land reform. As farmers acquired ownership of the land, agricultural production took off, stimulated by vast improvements in irrigation, new seeds, and support from an extensive farmers' association system. Taiwan was one of the earliest participants in the Green Revolution. Industrialization, initiated in the latter part of the Japanese era, was pushed aggressively also, focusing first on import substitution consumer goods, especially food processing. Sugar and other agricultural products in the food processing industry were the main exports at this time. Taiwan already was beginning a thrust toward export promotion, recognizing the limitations of the small domestic market, and the need to pay for necessary imports. In terms of the Rostow model of development, Taiwan could be described as in phase two, *Preconditions for Takeoff*. A major assist to this whole process came from the US, through financial aid to the government, military aid and protection, opening of American markets to Taiwan's exports, and active assistance especially to the agricultural sector through the Joint Commission on Rural Reconstruction, an American/Chinese joint operation, putting the knowledge and resources of American agriculture at the service of the ROC government and farmers on Taiwan. It was a highly successful partnership.

The 1950s also saw the start of multi-year development plans, which were guides developed by the government to help push the economy in the directions it felt best for the island (in stark contrast to the plans that ruled the command economy of the PRC on the mainland in this same period). Taiwan thus followed a path closer to the semi-planned economies that most other nations in the post-World War II era adopted, as opposed to the free-market, largely unplanned economic system in the US. Taiwan had six four-year plans, starting in 1953 and ending in 1976, after which it switched to six-year plans. The current plan, begun in 2002, is known as 'Challenge 2008' and is focused around measures that the government hopes will transform Taiwan into a 'green silicon island'.

1960s

The 1960s witnessed the continued strong growth of agriculture. But this was also the decade in which manufacturing overtook agriculture in its relative importance in the economy. Import substitution continued, but export expansion flourished even more, into various consumer goods, especially textiles, toys, and similar things. Taiwan's main asset was abundant low-cost labour that was starting to be drawn from rural areas into industrializing urban areas, as well as to rural-based manufacturing sites. Taiwan was somewhat unusual in Asia in this period in having a generally privately owned, small-scale industrial sector emerge and eventually become dominant in terms of number of workers employed and share of the island's total output. The native Taiwanese excelled at business entrepreneurship (a cultural characteristic that Chinese everywhere seem to share). Export Processing Zones (EPZs), among the first in Asia, were also established in Kaohsiung and other places, drawing together foreign capital and technology with local labour and prepared industrial sites. These EPZs were a natural expression of economic complementarity and were to become a standard mode of industrialization throughout much of Asia in succeeding decades. The Vietnam War was also a big help to the local economy in Taiwan, because the island provided much of the material needed by American forces fighting in Indochina, and was a key R&R stop for American forces as well. Within the Rostow model, Taiwan experienced the third phase, *Take-Off*, by the middle of the decade, coinciding with the withdrawal of American economic aid, which was felt to be no longer needed (military aid, however, continued).

1970s

In the 1970s, Taiwan's economy became more sophisticated, with the development of capital- and technology-intensive industries, including iron and steel, shipbuilding, petrochemicals, motor vehicles, and others, along with continuation of strength in consumer-goods industries. The growing wealth of the island was increasingly being invested in infrastructure also, especially electricity generation, telecommunications, and transportation. Expressways began to emerge, as private ownership of motor vehicles mushroomed (starting with motorcycles, and gradually phasing into automobiles). The train system expanded, with electrification and modernization, as it attempted to compete with highway transport.[29] Cross-island highways opened up not only the eastern part of the island to greater settlement and development, integrating it more and more into the island's economy, but also opened the interior mountains to unprecedented exploitation.

1980s

In the 1980s, Taiwan passed into phase four of the Rostow model, the *Drive to Maturity*, as strong economic growth in other Asian countries put increased pressure on Taiwan to constantly improve efficiency and move on to higher value-added manufactures that made even better use of Taiwan's highly educated and increasingly urban population. This pressure on Taiwan especially began to come from China, which started to abandon Maoism at the end of the 1970s, and was now attempting to emulate the already successful path followed by Japan first, then the four newly industrialized economies (NIEs)[30] of South Korea, Taiwan, Hong Kong, and Singapore – the 'Four Tigers', or the 'Mini-Dragons', as they also were dubbed. Hence, Taiwan moved now into consumer electronics, information technology and products. Taiwan became a leading producer of computer hardware and peripherals, among other things. The Hsinchu Science-Based Industrial Park played a key role in this upgrading of the island's manufacturing sector.

Toward the end of the decade, Taiwan underwent a political revolution that brought an end to some 40 years of repression, and began a renaissance in the political/human rights arena that had already been experienced in the economic arena. The watershed year was 1987, when martial law formally came to an end, and opposition parties began to flourish. The late 1980s thus not only marked a political watershed for Taiwan, but also ushered in the island's fourth technological and social transformation.

Taiwan's fourth technological and social transformation

1990s

By the 1990s, as the US and much of the world economy boomed in the high-tech bubble (even while Japan sank into a decade and more of stagnation in its economy), Taiwan became ever richer and moved into Rostow's fifth phase, *High Mass Consumption*, with consumers enjoying ample disposable income and indulging in material consumption (and foreign travel) at record levels. By most international standards, Taiwan was now a fully developed economy, with a per capita income equal to that of many European countries. Yet, the island was increasingly feeling the pressure of a number of challenges: appreciation of the NT dollar, wage rises, labour shortages, international trade competition, and growing demands for environmental protection (see later sections of this volume). China became a formidable economic power in the world economy in this decade, putting ever more pressure on Taiwan. To stay competitive, especially with labour costs, Taiwan found it more and more necessary to resort to 'offshoring' or 'outsourcing', as the buzzwords became known, i.e. relocating manufacturing operations to lower-cost sites outside of Taiwan, while retaining headquarter operations in Taiwan.

Geography and culture, as well as abundant, cheap labour, made China especially attractive as an investment site. Politics made China possible as a site for investment, when the Taiwan government lifted some restrictions on contact between Taiwan and the PRC. Thus, from the late 1980s onward, Taiwanese businesses started to invest tens of millions of dollars in China, especially the newly established Special Economic Zones (SEZs, the PRC's version of Taiwan's EPZs) in Xiamen, Shenzhen, and others, as well as into other parts of Guangdong, Fujian, and Jiangsu provinces, and the municipality of Shanghai, China's premier city and industrial centre. Investment and trade soared. Approved investment by Taiwan companies in China amounted officially to over US$20 billion from 1991 to March, 2002, making Taiwan the fifth largest foreign investor in China.[31] China has benefited enormously from this cash and technology flow. Significant investment from Taiwan also was sunk into Southeast Asia, especially the Philippines and Vietnam, but China dominated in its share of Taiwan's overseas investment. The situation was filled with irony. Taiwan and the PRC remained deadlocked in their political struggle, yet Taiwan was becoming ever more dependent on China for cheap labour and other production costs, and for export markets.

Thus, as Taiwan moved into the first decade of the twenty-first century, with a wealthy, consuming, overwhelmingly urban society of nearly 23 million people, the dominating theme became: How could Taiwan maintain prosperity and continue to grow, while struggling to establish a viable democratic system, all the while faced with overwhelming pressure from China for reunification and being drawn ever deeper into economic dependence on the mainland? The economic integration of Taiwan with the PRC continues unabated. Taiwan's trade with the PRC has been on a steadily steeper upward curve since 2001, with exports of $44 billion to China in 2005, and imports of $20 billion from the PRC. China now is Taiwan's biggest export market, well ahead of the $29 billion sent to the US, formerly Taiwan's largest overseas market. If one adds in Taiwan's exports of $34 billion to Hong Kong, much of which are then re-exported into China, then China and Hong Kong together account for almost 40 per cent of Taiwan's total exports.[32] Taiwan's annual trade surplus would cease to exist if it were not for the China trade. Indeed, Taiwan's economy could not survive now without China (and Hong Kong).

The relationship with China thus has become one of the main determinants of Taiwan's future, economically as well as politically. Opinions on the island are deeply divided as to whether Taiwan should integrate even more with the PRC or try to pull back. Some observers have speculated that eventually Taiwan might fall automatically into the PRC's lap, like a ripe apple from a tree, because of the island's dependence on the mainland economy. Thus, economic decisions of long-term impact hinge on this ongoing debate (see Chapter 5).

Role of politics

Closely intertwined with the economic transformation of Taiwan over the past 50 years has been the political system in Taiwan. Politics influences the economy in every country, every society. Nonetheless, Taiwan has undergone some dramatic political changes that have had profound impact on the economy, and hence on the environment.

As noted earlier, Taiwan's evolution since the Japanese colonial era is divided into two distinct periods separated by the watershed year of 1987. Prior to that, the island was governed under a single-party authoritarian system, defined as 'Leninist' by some, which was not all that different from the single-party communist system under Mao on the mainland.[33] The ROC government, under Chiang Kai-shek and the KMT, had several objectives. The first was to stamp out all political opposition, under what became known as the 'White Terror', in which dissidents were executed or imprisoned under a perverted justice system, the rationale being that Taiwan could not afford political opposition in the face of the dire threat from the Communists across the Taiwan Strait. A second objective was to carry out a determined educational and propaganda campaign to re-sinify, or assimilate, the Taiwanese and restore their true 'Chineseness', in the belief by the KMT that the Taiwanese had been so Japanized during 50 years of colonial rule that their loyalty to the ROC could not be trusted. Part of this initiative involved promoting Mandarin Chinese as the 'national' language, and suppressing Taiwanese language and culture (just as the Japanese had done, especially in the latter years of their rule). The irony of both these objectives is that in many ways they had the unintended consequence of reinforcing or encouraging a sense of Taiwanese identity and strengthening Taiwanese antipathy toward the Mainlanders and the KMT. A third objective was to use the stranglehold on the political system to ram through development policies that the government felt were essential to build up the economic (and military) strength of Taiwan, so that Taiwan would be able to at least defend itself against the threats coming from the PRC, and eventually to somehow retake the mainland from the Communists.[34]

From an environmental viewpoint, the consequences of the political system during the martial law era were profound. In the absence of public control over government decisions, in a system lacking normal checks and balances, the government basically could do whatever it wanted, without worrying about the views of special interest groups, or even public welfare in general.[35] There were no tedious environmental impact assessments that had to be followed, no public hearings, no consulting with opposition political parties to reach compromises. Whatever the government wished to do, it did, even if it knew there might be negative environmental consequences. The highest priority was the economy and defence, and the environment would just have to wait. The people of Taiwan were not totally innocent bystanders in this process, however. While it is true that engaging in public protests or showing opposition of any kind toward the government and its policies entailed high risks (and, hence, few dared do so), nonetheless, the people of Taiwan were 'bought off', in a manner of speaking, by the strong growth in the economy, the rapidly rising per capita incomes and consumer spending, and the increasing prosperity of the island in general. Most people on Taiwan were preoccupied with getting ahead economically for their individual selves and their families. Politics was a subject to be avoided. The environment and public welfare were peripheral issues during most of this era, for the government and for the citizenry.[36] The top priority was national security and economic growth, followed by regional development and redistribution of wealth. Environmental quality was in a distinctly third place. As a result, since 1987 developing a sense of public consciousness and concern by individuals for the environment and public good has been a real challenge for the government, for NGOs, and activists in the environmental movement.

The consequences of this situation are detailed in succeeding chapters of this volume, but a few illustrations will suffice here. One consequence was the deterioration in environmental quality in the rapidly expanding cities, especially Taipei and Kaohsiung (Figures 1.8 and 1.1). The artificial political governance of Taiwan meant that Taipei was made the 'temporary national capital' of the 'Republic of China', with all the associated functions and structures. As a result, it exploded in population. Today the Taipei-Keelung Greater Metropolitan Area has 6.6 million people and is the key economic, political, and cultural centre on the island. Taipei has become one of the key metropolitan centres of Asia's rapidly globalizing economy. Second in importance is the Kaohsiung Greater

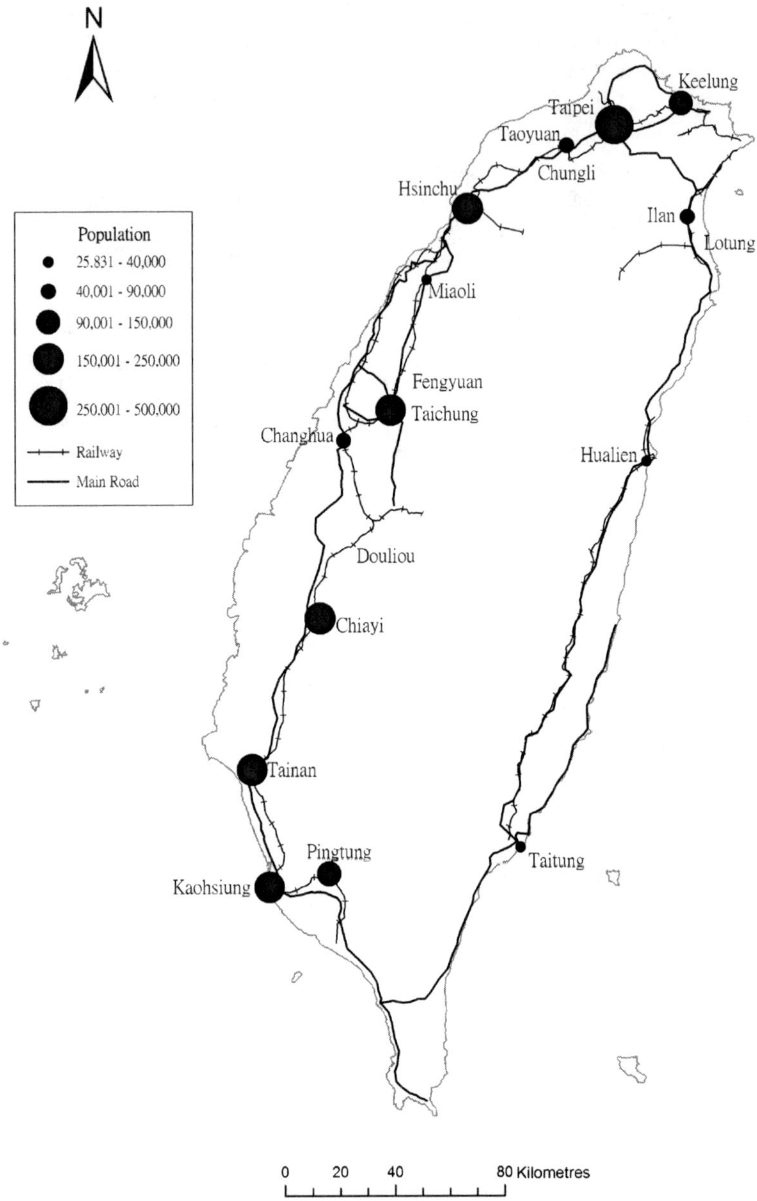

Figure 1.8 Taiwan: urban and transportation patterns, 1950 (source: Chen Cheng-hsiang, *Taiwan's Population* (1951)).

Metropolitan Area with 2.7 million residents. Third is the Taichung-Changhua Greater Metropolitan Area at 2.1 million population. The main urban areas account for nearly 70 per cent of the population, with more than 42 per cent alone concentrated in the Taipei region. As a result of this rapid urbanization, motor vehicle traffic grew enormously, resulting in some of the worst air pollution in the world. Taipei's wastewater and sewage system was hopelessly inadequate to handle the population, leading to severe water pollution in the rivers in the Taipei basin, which became open sewers. The rapidly expanding population also generated immense volumes of solid trash, which was collected in an antiquated and environmentally unfriendly way, undermining public health. In Kaohsiung, all of these conditions also developed. But there the air and water pollution were made even worse by being the site of much of the island's most polluting heavy industry – especially iron and steel, and petrochemicals. Kaohsiung earned a terrible reputation as just about the worst place in Taiwan to live, a stereotyping that the city still struggles to overcome.[37]

As a result of this rapid industrial and urban development, Taiwan has a very real environmental burden, compared with most other states in the world. Table 1.2 compares Taiwan's position with selected other states as of 2001 in terms of the density of population, factories, vehicles, and energy consumption.

In its efforts to develop industrialization, and promote private enterprise among the native Taiwanese (and hence increase their economic well-being and divert attention from political opposition to the KMT-controlled government), rural industrialization was promoted throughout much of the west coast plain. Thousands of small and medium-sized plants sprouted all over, in rural areas and small towns, and often with minimal or non-existent environmental controls. This rural industrial diversification policy was ended in the early 1980s, and the government then switched to promoting the development of industrial parks as the best way to promote economic growth. However, many of these industrial

Table 1.2 Taiwan's environmental burden in comparison with other selected countries, 2001

	Population density (pop./km²)	*Factory (factories/km²)*	*Vehicle density (vehicles/km²)*	*Energy density consumption*[1]
Taiwan	611	2.78	453	2,149
Japan	334	1.14	219	1,340
Germany	230	0.12	126	953
UK	238	0.06	102	230
USA	27	0.04	21	219
Taiwan/other countries (times)	1.6–23	2.4–69.5	2–21.5	1.6–9.8
Taiwan's world rank	2	1	1	1

Source: www.epa.gov.tw/English/offices/f/blueskybluesky2.htm.

Note

1 (tons equivalent fuel/km²).

parks were sited on fragile coastal lands, primarily because that was where land was most cheaply available in crowded Taiwan (Figure 1.9). The result was poisoning of agricultural soils in many parts of the island, and pollution of many of the island's rivers and streams (some severely). Heavy metals and other toxic wastes were dumped or buried here and there, like hidden time bombs.

Pressure to provide adequate incomes for the island's dwindling number of farmers, who faced a diminishing supply of good agricultural land as urban/industrial expansion continued virtually unchecked, led to various government schemes to promote cultivation of mountain slopelands. The main target was that intermediate zone between lowlands and the high mountain interior, where the government argued that proper development of terraces and use of suitable agronomic techniques could provide much valuable land for fruit cultivation and other crops.[38] Tens of thousands of hectares were opened up in this way, contributing without doubt to the agricultural economy and the food supply of the citizenry. Many farmers were already illegally and improperly using the slopelands for such activities, it was argued, so the government was simply trying to bring order to the system. Nonetheless, critics questioned the scale of the endeavour and some of the project locations and methods, in relation to soil erosion problems and siltation of reservoirs and rivers. Moreover, slopeland cultivation was sometimes extended into the high mountain country, alongside cross-island highways that the government built. Much of this activity was associated with the controversial programme designed to provide jobs and support for former ROC servicemen, mostly Mainlanders, through the Vocational Assistance Commission for Retired Servicemen (VACRS). The biggest project was in the Upper Tachia River Basin along the first cross-island highway and this is discussed in more detail in Chapters 2 and 4. Suffice to note here that the programme proved highly controversial and had profound consequences for the mountain interior.

Related to that were KMT policies toward the aborigines, especially those who lived in mountain villages and struggled to adapt to Taiwan's rapidly changing economy. Like the Native Americans, or 'Indians', in the US, the aborigines in Taiwan were a neglected minority living at the bottom of the socioeconomic ladder, largely ignored and unknown by Taiwan society, with high incidence of alcoholism, juvenile delinquency, unemployment, health problems, and other social pathologies. Their mountain realm was being invaded more and more by lowlanders, especially Mainlanders in control of everything, and many of the aborigines felt they were not getting a fair deal from the ROC government, which greatly restricted access by lowlanders (or foreign visitors to Taiwan, for that matter) to aboriginal areas. Ostensibly this was to 'protect' the aborigines and their mountain environment; critics claimed it was to hide the neglect and abuse of the aborigines and their lands by the KMT government.[39]

Another highly controversial government policy during the KMT era was development of nuclear power plants for electricity generation. With virtually no energy resources except for hydro power, which could never begin to meet more than a small fraction of the island's mushrooming energy demands, the ROC

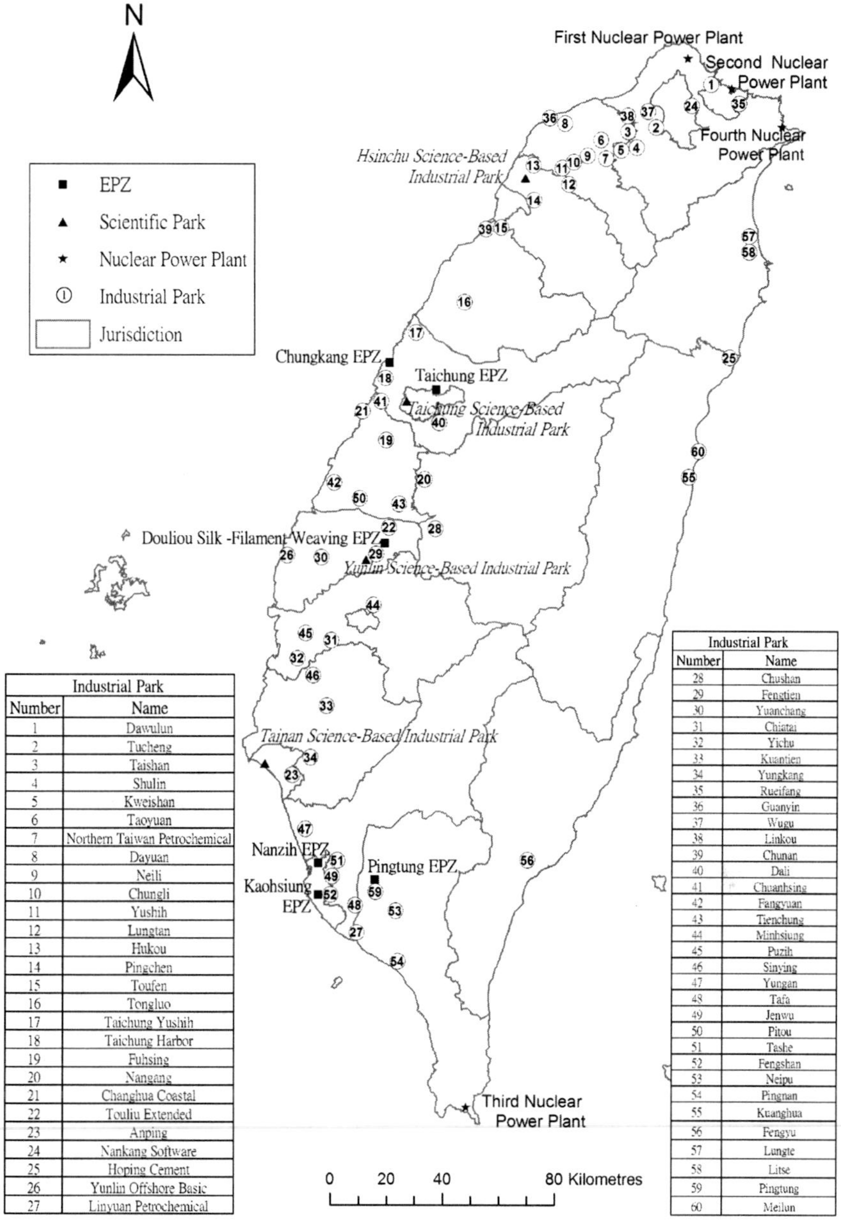

Industrial Park	
Number	Name
1	Dawulun
2	Tucheng
3	Taishan
4	Shulin
5	Kweishan
6	Taoyuan
7	Northern Taiwan Petrochemical
8	Dayuan
9	Neili
10	Chungli
11	Yushih
12	Lungtan
13	Hukou
14	Pingchen
15	Toufen
16	Tongluo
17	Taichung Yushih
18	Taichung Harbor
19	Fuhsing
20	Nangang
21	Changhua Coastal
22	Touliu Extended
23	Anping
24	Nankang Software
25	Hoping Cement
26	Yunlin Offshore Basic
27	Linyuan Petrochemical

Industrial Park	
Number	Name
28	Chushan
29	Fengtien
30	Yuanchang
31	Chiatai
32	Yichu
33	Kuantien
34	Yungkang
35	Rueifang
36	Guanyin
37	Wugu
38	Linkou
39	Chunan
40	Dali
41	Chuanhsing
42	Fangyuan
43	Tienchung
44	Minhsiung
45	Puzih
46	Sinying
47	Yungan
48	Tafa
49	Jenwu
50	Pitou
51	Tashe
52	Fengshan
53	Neipu
54	Pingnan
55	Kuanghua
56	Fengyu
57	Lungte
58	Litse
59	Pingtung
60	Meilun

Figure 1.9 Taiwan's coastal development projects (source: Industrial Development Bureau, Ministry of Economic Affairs, 2006).

government claimed it had no choice but to push for nuclear power, in order to reduce the expensive importation of coal and oil for power plants (plus the environmental consequences of coal burning). Hence, over the years a total of three nuclear power stations (containing six nuclear power units) were constructed. Plans for a fourth plant were started in 1980, but the project became embroiled in controversy because of anti-nuclear activists. Nuclear power provided about 8 per cent of Taiwan's total energy needs in 2003.[40] The long-range intent of the former KMT government was to expand this aggressively, in spite of growing public unease about the safety of the plants in such a high population density island, aware of such disasters as those at Three Mile Island in the US and Chernobyl in the Soviet Union. Would Taiwan have developed nuclear power in this way, or at all, if the political system had been more democratic during this era? Perhaps not. In any event, the DPP government declared its intention to eventually eliminate nuclear power from Taiwan.

Things started to change, dramatically, with the lifting of martial law in 1987. The most fundamental change was to open the doors to legal political dissent and opposition parties, the most prominent being the Democratic Progressive Party (DPP), which had actually been in the formative stages for a number of years. This change was made possible partly through the efforts of President Chiang Ching-kuo (CCK), who succeeded his father, Chiang Kai-shek, in 1978.[41] Lee Teng-hui, a KMT member but native Taiwanese, came to office in 1988 when CCK died. Lee pushed the democratization process much further. Now, everything was out in the open, everything became possible, and Taiwan blossomed remarkably quickly into a vibrant democracy.[42] A free press also bloomed, along with a multitude of non-governmental organizations (NGOs), many of them dedicated to various environmental and civil rights causes.

One momentous change, already alluded to, was to reopen contacts with the mainland, as a result of changes in the economy and the need to find sources of cheaper labour and other production inputs. Thus, the ROC government, starting in the mid-1980s, but especially accelerating with the 1990s and beyond, made it legal for the people of Taiwan to develop economic and cultural/social linkages with the mainland, as long as they were developed indirectly, without the direct involvement of the national government.[43] Thus, trade with and investment in the mainland both mushroomed, as already noted, as did tourism by visitors from Taiwan. Hong Kong became the main conduit for these growing ties between Taiwan and the PRC, and the short Hong Kong–Taiwan air route became one of the most heavily used in the world as jumbo jets ferried millions of people back and forth each year. Publicly, the ROC government clung tenaciously to the strict Three No's policy (no direct contact, no compromise, no negotiation) that had been the cornerstone of their policy toward the PRC since 1949, in spite of increasing pressure from business people especially to abandon this artificial and untenable position. However, in reality, the ROC was in fact developing growing ties, including contact and negotiations through semi-private organizations that were set up to handle the growing relations between the two sides: the Straits Exchange Foundation (SEF) representing Taiwan, and the Association for Relations Across the Taiwan Strait (ARATS) representing the PRC. Theoretically, these two so-called non-government organizations enabled the ROC government to maintain the fiction that it was having nothing directly to do with the PRC.[44]

One of the advantages gained by Taiwan businesses in the growing volume of Taiwan-owned operations in the PRC was that the PRC tended to have even more lax environmental rules and regulations than in Taiwan, in addition to much lower labour costs. Thus, the investment environment in the mainland was irresistible for Taiwan companies, and today most Taiwan companies have some kind of operation going in the mainland.

Conclusion

One of the greatest benefits of the new post-martial law era was establishment of a true environmental movement, and the founding of the government's Environmental Protection Administration (EPA).[45] By the 1980s, discontent was bubbling in various parts of Taiwan over environmental abuses and the lack of accountability by government and private businesses, especially those enterprises creating havoc with the environment. Various incidents started to take place in locations such as Lukang, Houchin, Neihu, Linyuan, Lanyu, Hungmao, and others, in which local citizens became activists because of policy decisions and consequences they did not like and were no longer willing to be passive about.[46] The environmental movement was being born. With the lifting of martial law, NGOs began to sprout, many dedicated to various facets of environmental protection and improvement. In a classic bottom-up story, citizen activism forced the government to get serious about environmental protection, in a pattern remarkably similar to Japan's experience in the 1960s and 1970s. This led to the formal establishment of the cabinet-level EPA in August, 1987. The next year, the provincial government set up its own environmental protection office, as did other local government units around the island. Environmentalism became a politically popular subject. In the years since that watershed development, Taiwan has come a long way in attacking environmental problems resulting from some 40 years of largely unregulated growth. This story is covered in succeeding chapters, but suffice here to note that the road has been rocky, the problems (political, economic, social) many, and the accomplishments uneven. Nonetheless, it is safe to say that Taiwan today is definitely a cleaner, healthier place to live than it was 20 or more years ago. There is something to be learned from Taiwan's experience.

2 The perils of development

Taiwan's environmental problems and their causes[1]

Taiwan has made enormous economic and material progress since the 1950s, not to mention striking social and political gains since the late 1980s, as was noted in Chapter 1. Yet, if one were to poll a random sample of people in Taiwan today, a majority probably would say that Taiwan is 'badly polluted', or words to that effect. A survey by the Environmental Quality Protection Foundation in 2006 reported that public frustration with the quality of the environment registered a high of 70 'points', up slightly from the previous year (the index peaked at 79 points in 1995, then decreased before starting to rise again). The survey also noted that 37 per cent of the people felt that Taiwan's environment had actually deteriorated over the past five years, while 24 per cent thought it was about the same.[2] Hardly a ringing endorsement of the government's efforts at environmental improvement.

People have different perceptions, though, of what constitutes the natural 'environment' or what 'pollution' means to them.[3] For some, awareness of the environment is limited to those elements that directly impact their daily lives, such as dirty air and water, garbage, etc. Those elements of the environment that are out of sight also tend to be out of mind. At the same time, most people probably would say that the quality of life in Taiwan today is much better than before. But 'quality of life' is also a fuzzy concept that means different things to different people. Ideally, it should include a great many things, such as the state of the economy, education, culture, recreation, social welfare, medical conditions and care, public order, government, and, of course, environment.[4] None of these sectors is ideal or utopian in Taiwan (nor in any other place in the world, for that matter), but it is failings in the last category, environment, that probably pulls down the quality of life the most at the present time.[5]

One perspective on Taiwan's environmental standing comes from an outside source, as reported by the EQPF of Taiwan. The Environmental Sustainablity Index (ESI), produced by a consortium of Yale, Columbia and centres, uses some 20 'core indicators' of environmental quality to do an annual ranking of a majority of the world's nations. Taiwan is not formally included in the ranking (because of its non-nation status and not being a member of the UN), but some scholars in Taiwan calculated Taiwan's standing in the ESI on an ad hoc basis, and determined that Taiwan would have ranked 72nd out of 143 countries in

2002 (vs. 42nd out of 123 countries in 2001).[6] While this ESI, like all indices, is subjective and flawed, nonetheless, it does suggest that Taiwan still has a long road to go in its quest for environmental quality.

In another international comparison, the World Economic Forum's (WEF) *2006 Environmental Performance Index* (EPI), Taiwan came up 24th out of 133 nations included in the study. Taiwan trailed behind most of Western Europe and Japan, but interestingly was slightly ahead of the US (28th), as well as Russia and South Korea. The EPI focuses on a nation's environmental performance within the context of sustainability, and provides a means of showing the relationship between economic competitiveness and environmental protection. Specifically, the EPI uses 16 indicators tracked in six policy categories (environmental health, air quality, water resources, biodiversity and habitat, productive natural resources, and sustainable energy). Taiwan's standing in this ranking thus was relatively good. Observers in Taiwan have often complained that international rankings commonly use incomplete or erroneous data for Taiwan, perhaps because of Taiwan's exclusion from the UN and most other international organizations, and thus the island tends to get unfair representation in the world. Hence, this 2006 EPI was better received in Taiwan, although it did not gloss over Taiwan's problems. Yeh Jiunn-rong, Executive Director of Taiwan's National Council for Sustainable Development, said the EPI report honestly reflected Taiwan's problems, such as over-fishing, poorly designed agricultural subsidies, low use of renewable energy, and others.[7]

Regardless, the contrast between these two international comparisons, the EPI and the ESI, does suggest that these kinds of surveys, based on macro data of sometimes different dates and varying reliability, should not be taken too literally, but seen only as broad approximations of the relative state of the environment.

Perhaps the most ringing indictment, though, of Taiwan's environmental quality comes from the government's own National Council for Sustainable Development (see Chapter 3 for discussion of this organization), which notes in its 2006 Annual Report:[8]

> Index scores for environmental pollution have shown slight fluctuation below the 100 mark from 1988 to 2005. [Note: 1988 is used as the index year, just after the TEPA was established.] … over the long-term, environmental quality has not noticeably improved. As for individual indicators, waste resource recycling rates are following a marked trend of improvement. However, only limited improvements have been observed in indicators for PSI average and reservoir water quality. The indicator for carbon dioxide emissions is on a downward trend showing emissions are not yet effectively under control.

For the purposes of this study, the 'environment' is interpreted in its broadest possible context, to include all facets of the land, air, and water that directly or indirectly impact the quality of life of human beings, whether living in urban or rural areas, and their ability to carry out their occupations and accustomed lifestyles.[9] Hence, we are examining here the total ecological system in which Taiwan exists. While our attention is primarily focused on the internal environmental problems of Taiwan, we also need to examine Taiwan's role or position within the context of regional and global environmental issues, such as acid rain, ocean water pollution, the 'greenhouse effect' and global warming, and others. This chapter thus is a kind of inventory of the host of environmental problems that afflict Taiwan at the present time and the factors that cause those problems.

Humans live in ecosystems consisting of land, air, and water, in which fauna and flora co-exist (or try to). Without question, humans are the dominant species on the planet, and hence wherever humans are located they have a profound impact on natural ecosystems and anything else, plant or animal, trying to live there. Thus, in breaking down the environment into its component parts, it is perhaps easiest to approach it first from the perspective of the impact of wastes produced by human society. In simplest terms, those wastes consist of solids, liquids, and gases. In rapidly expanding economies, production of these wastes tends to grow much faster than increases in population, simply because per capita consumption rates (and hence production of wastes) tend to increase exponentially as humans pursue high mass consumption of goods and services. Virtually the whole world economy is geared to this paradigm. Even the poorest of societies aspire to this model, even if they have not yet achieved it. Taiwan is one of the states that succeeded in attaining this goal of high mass consumption in recent decades. One consequence, however, is large volumes of wastes, which too often are poorly handled and end up polluting the air, land, and water, and ultimately affect human health and well-being. The cumulative effect is deterioration in the quality of life, even as material consumption rises.

Types of human waste

More specifically, human wastes can be divided into four major categories or types: inert and semi-inert substances; biologically active wastes; hazardous wastes; and atmospheric emissions.

Inert and semi-inert substances

Inert and semi-inert substances – glass, plastics, metal, etc., found in ordinary garbage – are materials that are basically harmless in and of themselves, but cause problems because of their bulk and the fact that they do not biologically degrade and hence survive almost indefinitely. Especially in modern urban societies, these are among the most visible and common wastes, often mixed with organic or toxic substances. These substances tend to be bulky and hence the volume can become enormous, resulting in unsightly accumulations that also can pose health hazards to humans, as well as to wildlife that ingests them or gets tangled in them, especially when the waste materials are piled up near rivers or lakes and the substances somehow get into the water.

Biologically active wastes

Biologically active wastes are those that will eventually biodegrade. These include food wastes, paper, fertilizers, human and animal bodily wastes, and carcasses. The main problem with these wastes is what to do with them while waiting for them to biodegrade, which can take quite some time if allowed to proceed at natural rates. While decomposing, these wastes can contribute directly and quickly to health problems, and hence disposing of these substances properly takes on an urgency not normally associated with inert wastes. These biologically active wastes also usually are highly visible, messy, and smelly, and hence ones that the public often complains about.

Hazardous wastes

Several notches higher in danger to society is a third group of wastes that are deemed outright hazardous. These include caustic chemicals, pesticides, radioactive wastes, heavy metals, and a wide range of other chemically potent substances used or produced by industry, agriculture, or consumers. Although potentially life-threatening, these wastes commonly are not immediately visible and may even be difficult to detect without special scientific equipment. Their hazardousness also can be extremely long-term, such as in the case of radioactive wastes. These are the most potent residue of modern industrial society and the nuclear age. The impact of hazardous wastes on human health is still not completely known, but the dangers they pose to all life on the planet have become increasingly studied in recent decades. Increased awareness of the problem in Asia was due in large part to the efforts of scholars and activists, through such publications as the shocking *Island of Dreams* in the 1970s, which brought to light the horrors of Japan's environmental crisis and struggle to deal with such things as heavy metal pollution,[10] or pioneering studies by Smil, He, and others in exposing the environmental degradation taking place in China.[11] Dealing with these kinds of hazardous wastes poses the greatest challenge to societies around the world, because the effort requires a high level of scientific knowledge, special equipment, much manpower, and much time. The cost can be enormous, as illustrated by such well-known past disasters as the effort to contain the contamination at Chernobyl in Ukraine, or the effort to clean up the Love Canal area in Buffalo, New York, to mention just two of the most infamous sites.

Atmospheric emissions

The three categories above involve solid and liquid wastes. Atmospheric emissions consist of gases of various kinds, such as carbon monoxide, hydrocarbons, oxides of nitrogen and sulphur, polyaromatic hydrocarbons, chlorofluorocarbons, and others. These gaseous wastes are the result of emissions by motor vehicles and factories, burning of waste materials, use of aerosols, and other actions. Whatever the source, the result is air pollution. This kind of pollution tends to be worst in urban areas, simply because that is where motor vehicles, factories, and people are concentrated. But the impact can extend well beyond urban areas, because of prevailing winds and weather patterns, which sometimes distribute the pollution over large expanses of rural or wilderness areas. The reach can be global, as in the case of acid rain or global warming. Hence, pollution of the air over Taiwan is not necessarily confined just to Taiwan.

Types of degradation

Whether the waste products are solid, liquid, or gas, and whether inert, biodegradable, or hazardous, the end result can be some kind of 'degradation' or 'pollution', somewhat vague terms that carry different meanings to different people. Of all the types of pollution, however, none probably are better known or more visible to the average citizen than air and water pollution, because people encounter these in their everyday lives, whether they want to know or not. But also part of the problem is noise, and pollution or degradation of land.

Air pollution

Air pollution may or may not be visible to the naked eye. But air that obscures visibility, or smells (or, even worse, tastes) bad, or causes breathing and health problems for people tends to generate public controversy and complaints to authorities. This kind of pollution is most noticeable in urban areas, where most people live in Taiwan today (as now is true in most of the world's societies). Scientists in Taiwan have adopted the Pollutant Standard Index (PSI) used by the US EPA to measure air quality.[12] The PSI value is an index indicating the daily highest concentration level of five pollutants: particulate matter of ten microns or less in diameter (PM_{10}), ozone (O_3), sulphur dioxide (SO_2), carbon monoxide (CO), and nitrogen dioxide (NO_2). A PSI value below 50 is considered 'good', 50 to 100 'moderate', and greater than 100 'unhealthful'. Values beyond 200 are outright dangerous (and fortunately are virtually unheard of today in Taiwan).[13] For Taiwan as a whole, air quality has been steadily improving since the late 1980s, looking at it just from the perspective of PSI values. The number of days per year with unhealthy air quality (i.e. PSI > 100) peaked at 16 per cent in 1991, but has hovered between 2.6 and 4.6 per cent in the last five years (Table 2.1). In the 1980s the main air pollutants causing PSI values greater than 100 were particulate matter, sulphur dioxide, carbon monoxide, and ozone, especially the first one. Particulate matter and ozone have since reversed positions, however, in their role in air pollution, from 76 per cent and 24 per cent, respectively, in 1994, to 22 per cent and 78 per cent in 2002, meaning that ozone has become the biggest headache.[14] Table 2.2 shows the changes in degree of air pollutants since 1994, and only ozone has increased, although even it has been fairly stable in the last five years. Sulphur dioxide and particulate matter have improved the most.

Table 2.1 Changes in PSI levels in Taiwan, 1985–2005

	PSI (ave.)[1]	*PSI 0–50 (%)*[2]	*PSI 51–100 (%)*	*PSI > 100 (%)*
1985	na	22.1	64.2	13.3
1990	83.8	18.5	65.2	16.3
1995	67.5	35.9	58.0	6.1
1996	64.5	39.7	53.8	6.6
1997	62.7	40.3	54.2	5.4
1998	58.2	47.4	47.6	5.0
1999	59.9	46.3	48.6	5.1
2000	59.4	46.3	48.7	5.1
2001	57.9	43.4	53.2	3.4
2002	54.3	42.2	54.6	3.2
2003	55.0	42.3	55.0	2.6
2004	62.1	38.6	56.8	4.6
2005	62.8	39.8	55.8	4.4

Source: *Yearbook of Environmental Protection Statistics, 2006*, pp. 2–2, 2–3, 2–16.

Notes
1 $\mu g/m^3$.
2 Per cent of days.

Table 2.2 Changes in air pollutants in Taiwan, 1989–2005

Year	*SO_2(ppm)*	*CO(ppm)*	*O_3(ppm)*	*PM_{10}(ug/m^3)*	*NO_2(ppm)*	*NMHC (ppm)*
1989	0.028	3.30	0.036	87.2	0.022	0.66
1990	0.027	3.26	0.042	83.8	0.025	0.84
1995	0.009	0.92	0.052	67.5	0.024	0.53
1996	0.007	0.96	0.054	64.6	0.024	0.58
1997	0.006	1.03	0.054	62.7	0.024	0.57
1998	0.006	0.81	0.053	58.2	0.023	0.41
1999	0.005	0.74	0.054	60.0	0.023	0.37
2000	0.004	0.70	0.054	59.4	0.022	0.30
2001	0.004	0.73	0.057	57.8	0.021	0.37
2002	0.004	0.60	0.060	54.3	0.019	0.26
2003	0.003	0.62	0.060	55.0	0.019	0.39
2004	0.004	0.55	0.061	62.1	0.020	0.36
2005	0.005	0.54	0.057	62.8	0.018	0.27

Source: *Yearbook of Environmental Protection Statistics, 2006*, pp. 2–16, 2–17.

Of course, these are averages for the whole island. For urban/industrial regions (Table 2.3), while the ten years from 1994–2003 saw notable improvement in days with unhealthy PSI levels, some regions have worse air than the island average, most notably the Kaohsiung/Pingtung heavy industrial region in the south. Long the area with the worst air pollution, nonetheless the KaoPing region has cut its bad air days by two-thirds.

Particulate matter smaller than 10 microns is a particularly insidious pollutant, because these particles are more likely to become lodged in lung tissue.

Table 2.3 Unhealthy PSI levels by urban/industrial regions in Taiwan, 1994 vs. 2003 (PSI > 100 in % of year)

	1994	*2003*	*Improvement (%)*
Northern	3.4	1.3	62
Hsinchu/Miaoli	2.1	0.7	67
Central (Taichung)	5.1	2.7	47
Yunlin/Chiayi	5.3	2.6	51
Kaohsiung/Pingtung	18.4	6.6	64
Hualien/Taitung	[None]	[None]	
Ilan	[None]	[None]	

Source: *Environmental Yearbook 2004*, p. 64.

Combined with ozone, as well as the other pollutants, this mixture can do real damage to the human body, depending on concentrations, length of exposure, and the age and general health of the person breathing the polluted air. It is not surprising that cancer has become the leading cause of death in both men and women in Taiwan, and lung cancer competes with liver cancer as the leading form of the disease for men, while liver cancer is the number one form for women.[15] The role of air pollution (and its components) in the incidence of lung cancer is not entirely known, but is under intensive study in Taiwan and elsewhere (heavy cigarette smoking in Taiwan is likely to be a contributing factor as well). More directly understood, however, is the link with bronchitis, emphysema, and asthma. In 1996, for example, one study revealed that more than 12 per cent of Taipei's high school students suffered from asthma, the highest incidence on the island.[16]

The major sources of air pollutants in 2003 can be seen in Table 2.2. In sheer volume, hydrocarbons are largest, followed by carbon monoxide, nitrogen oxides, suspended particles, and then sulphur oxides. Industry is the main contributor for sulphur oxides, nitrogen oxides, hydrocarbons, and lead, while construction accounts for the largest share of suspended particles, and motor vehicles for the vast majority of carbon monoxide. These sources and causes are understandable when seen in light of the more than 98,000 factories operating in Taiwan in 2003, many of these small-scale enterprises with relatively unsophisticated equipment. Also significant has been the rapid increase in number of motor vehicles. In 1989 there were 2.5 million cars and 6.7 million motorcycles in Taiwan, for a density of 258 per km^2. By 2003, those numbers had escalated to 6.1 million cars and 12.3 million motorcycles, for a density of 512 per sq. km^2. The density thus had nearly doubled in just 14 years[17] (Table 2.4).

The TEPA statistics may not tell the full story, however. Scholars working privately on pollution studies in Taiwan paint a less rosy picture of progress made and severity of problems with air (and other) pollution. For one thing, they point out that PSI air quality samples are taken at heights of 15 to 25 metres above ground level, an international practice, because samples taken closer to the ground fluctuate too much and thus are less reliable when measuring pollu-

Table 2.4 Environmental burden in Taiwan, 1981–2003

Year	*Population (10,000 persons)*	*Population density (pop/km²)*	*Motor vehicles (10,000)*	*Factories (1,000)*	*Factory density (per km²)*	*Pigs (1,000)*	*Energy consumed (mKWH)*	*Per cap. energy consump. (KWH)*	*Waste amount (10,000 tons/yr)*	*Daily waste per person (kg/yr)*
1981	1,813.55	503.76	541	60.28	1.67	4,826	–	–	356.28	0.63
1990	2,035.30	565.36	1,146	92.98	2.58	8,565	–	–	684.48	0.96
1995	2,130.42	592	5,841	97.01	2.69	10,509	36,437	1,718	870.77	1.14
1996	2,147.14	596	6,280	96.82	2.69	10,698	39,026	1,825	873.64	1.14
1997	2,168.33	602	6,688	99.85	2.77	7,967	40,003	1,854	888.80	1.14
1998	2,187.09	607	6,806	98.84	2.74	6,539	44,963	2,065	888.05	1.14
1999	2,203.41	612	7,030	100.68	2.80	7,243	44,761	2,039	856.57	1.08
2000	2,221.61	617	7,278	98.86	2.73	7,495	47,515	2,148	787.55	0.98
2001	2,033.98	620	7,315	97.21	2.69	7,165	48,634	2,148	725.48	0.90
2002	2,245.31	624	7,559	98.24	2.71	6,794	50,606	2,260	672.36	0.83
2003	2,253.48	626	7,777	98.87	2.73	6,779	53,175	2,364	613.91	0.75

Sources: 1. *Urban and Regional Development Statistics*, 2004, CEPD; 2. *Yearbook of Environmental Protection Statistics*, 2004, TEPA; 3. *Statement of Environment*, 2004, TEPA.

Note
All data are for end of year.

tion levels over a wide area. However, people moving around on the ground are not breathing air at 15 to 25 metres and hence may actually be inhaling much higher levels of pollutants than the PSI numbers suggest. Moreover, the PSI does not even include one of the most hazardous types of air pollution, known as volatile organic compounds (VOCs), of which benzene is one of the worst and was once widely found in the air of Taipei, for example.[18] These VOCs were present in the gasoline produced in Taiwan through the 1990s. The introduction of catalytic converters into automobiles, which started in 1990, has supposedly reduced this problem, but not yet eliminated it. Moreover, the more polluting two-stroke motor scooters once so common in Taiwan were banned by the end of 2003 in favour of less polluting four-stroke motorcycles, all with catalytic converters (see Chapter 3). A further complication with the PSI figures is that industrial air pollution is more hazardous than that from vehicular traffic. Thus, even though pollution from factory emissions has indeed fallen, in areas of highly concentrated, heavily polluting industry, such as parts of Kaohsiung in the south, unhealthy PSI levels are much more common than in places such as Taipei where the economy is more service-based (Table 2.3).

One growing source of pollutants for Taiwan's air is actually exogenous to the island, namely, dust and other air pollutants blowing over from China. In recent years, dust storms have been increasing in frequency and intensity in China. Taiwan has been affected by these dust storms every year since the early 1990s. Particularly serious instances occurred in March 1995, May 1996, March and April 2000, and April 2001.[19] The year 2005 started out as one of the worst also, with four storms blowing over Taiwan between March and May. According to the TEPA, the worst instance recorded to date, however, was on 12 March 1995, when the density of dust in the air was 100 times the norm.[20] During the worst of these storms, PSI levels in Taiwan have often risen as high as 171 (in 2000), but more typically are between 120–40. However, a storm in March 2006 brought PSI levels over 200 to Taiwan. The storm blew out of Mongolia and then swept east to Korea and Japan before turning south to blanket Taiwan. In the US, regulations allow a maximum safe level for a 24-hour period of 150.[21] Fortunately, these dust storms from China usually blow over within a few days at most and then air quality returns to normal.

Equally bad, if not worse, are emissions of sulphur dioxide coming from the thousands of power plants in China, which relies on coal to produce 80 per cent of its electricity. In 2003, China's total SO_2 emissions reached 21.6 million tons, second only to that of the USA. Carbon dioxide emissions are just as bad, having increased 45 per cent between 1990 and 2002. Hong Kong and Taiwan are affected especially by factories and power plants in neighbouring Guangdong province, China's economic powerhouse, while Korea and Japan are particularly impacted by plants in north China. Air pollutants from China have even been detected in the US mainland. The end result is the same, however – increased air pollution about which the receiving country can do very little.[22]

Water pollution

Taiwan's fresh water suffers various levels of pollution. Many of the rivers of Taiwan are polluted to some degree. Taiwan has a total of 129 rivers, 21 of them classified as primary, 29 as secondary. The TEPA has tested all of the rivers and, as of 2003, 15.9 per cent were heavily polluted, 11.3 per cent medium polluted, 13.4 per cent lightly polluted, while 59.4 per cent were non-polluted.[23] (Table 2.5) Unfortunately, these figures reveal a deterioration, not improvement, since 1990, with the biggest increase in proportion of rivers that are heavily polluted. The most abused rivers, understandably, were in the major metro areas, such as the Tanshui/Keelung/Hsintien system in the Taipei Basin.[24]

Groundwater in Taiwan also suffers, especially from very high rates of iron, manganese, and arsenic, with anywhere from 20–45 per cent of samples failing to meet water quality standards during the 1990s, and the failure rate rising to 70–80 per cent for iron and manganese in 2000 and 2001. Some samples also fail now because of presence of excessive chromium, cadmium, copper, and zinc. The data seem to suggest that groundwater is getting worse, not better.

Table 2.5 Levels of pollution in Taiwan's rivers, 1987–2005

Year		*No pollution*	*Light pollution*	*Medium pollution*	*Heavy pollution*	*Total*
1987	Length	1,998.9	180.9	319.7	321.1	2,820.6
	%	70.9	6.4	11.3	11.4	100
1995	Length	1,882.6	337.9	319.6	394	2,934.1
	%	64.2	11.5	10.9	13.4	100
1996	Length	1,816.1	302.4	358.6	434.6	2,911.7
	%	62.4	10.4	12.3	14.9	100
1997	Length	1,890.8	296.2	395	352	2,934
	%	64.4	10.1	13.5	12.0	100
1998	Length	1,887.9	272.4	442.3	331.4	2,934
	%	64.3	9.3	15.1	11.3	100
1999	Length	1,942.9	222.8	416.4	351.9	2,934
	%	66.2	7.6	14.2	12.0	100
2000	Length	1,865.3	353.1	360.9	354.8	2,934.1
	%	63.6	12.0	12.3	12.1	100
2001	Length	1,808.9	287.6	451.3	386.2	2,934
	%	61.7	9.8	15.4	13.2	100
2002	Length	1,812.5	349.7	335.4	406.5	2,904.1
	%	62.4	12.0	11.5	14.0	100
2003	Length	1,726.1	389.9	328.9	459.2	2,904.1
	%	59.4	13.4	11.3	15.8	100
2004	Length	1,860.1	284.8	537.4	222.0	2,904.2
	%	64.1	9.8	18.5	7.6	100
2005	Length	1,864.9	287.8	572.5	179.1	2,904.2
	%	64.2	9.9	19.7	6.2	100

Source: *Yearbook of Environmental Protection Statistics, 2006*, pp. 2–54, 2–55.

Note
All lengths in km.

Solid waste (garbage)

Water pollution is caused by many things, but solid waste and sewage are leading culprits. Take the case of solid waste, which can contain all three kinds of wastes: inert, biodegradable, and even hazardous (in the absence of any recycling effort, although hazardous materials are predominantly found in industrial waste, not ordinary household garbage). In Taiwan, 'solid waste' technically includes general waste, industrial waste, and hazardous waste, but we are focusing primarily on just general waste (garbage) in this subsection. Societies everywhere struggle to get control of this pervasive problem, which has several dimensions. One is in how to collect the substances in the first place. Littering of the landscape with this detritus of modern civilization is all too common. Most societies try to encourage members to dispose of these things in proper receptacles, or at least to put them in piles at centralized locations. Most societies then try to collect the materials in some organized way and haul them off to dumps or landfills. Some societies use incineration as a way to dispose of the huge bulk of these substances and thereby reduce or eliminate the health hazards. Some societies try to encourage use of alternative materials, especially biodegradable substitutes that will be easier to get rid of. Some societies encourage recycling. The methods and efforts are highly varied around the world. No society, rich or poor, has yet been entirely successful in solving the problem. Taiwan's experience in this regard is hardly unique.

Taiwan's output of solid waste (garbage) peaked at the end of the 1990s, according to the TEPA, at about 1.1 kg/day/cap., or roughly 400 kg/year per person.[25] With 22.5 million people, that amounted to a total of nearly 25,000 metric tons per day for the whole island's population or about nine million tons of garbage per year.[26] The per capita figure has since dropped below 1 kg/day (due to environmental education efforts, as well as householders now having to purchase garbage bags in Taipei and some other places), which is a positive trend. Sorting of garbage into categories, for recycling, has increased to about one-third of all solid waste, but is still not universal on the island (see Chapter 3). The TEPA has determined that the composition for 2003 was roughly 33 per cent paper, 27 per cent food wastes, 21 per cent plastics, and the remainder consisting of textiles, garden trimmings, leather and rubber, metal, glass, and others.[27] Regardless, the cumulative volume is considerable and highly visible.

One of the most troublesome items in the inert category has been the ubiquitous plastic bag, which is cheap, light, easy to store, and hence widely used on the island, especially in the retail sector. (This is true throughout Asia, for that matter.) Unlike in the United States, paper bags are much less commonly used to hold commodities purchased in retail stores, such as grocery stores or supermarkets.[28] Until just recently, it was estimated that Taiwan used 2.5 plastic bags per day per person, adding up to 105,000 tons or 20 billion bags each year, most of which ended up in garbage containers and piles.[29] Indeed, plastic bags used to be the most common receptacle used by households to wrap up garbage. Under the old collection system, these filled plastic bags (usually bags used first to carry retail purchases home) were then dumped by each household in piles at designated collection points in each neighbourhood, for pick-up later by garbage trucks. Unfortunately, the bags often split upon dumping, or stray dogs and rats chewed on them, opening the contents to the air and insects such as roaches. Rains might fall before the trucks could complete their rounds, resulting in food wastes and other trash being washed into the sewer system. Also part of the problem are Styrofoam and plastic food containers, widely used by the estimated 18 million people in Taiwan who eat out every day, creating an annual total of 59,000 tons of disposable tableware garbage. Once collected under the old system, the garbage (of whatever composition) was hauled off to some 300 or so landfill sites, many of which were full to capacity. Much garbage did not make it even to landfills but simply was dumped in rivers or in remote gullies. These piles of rotting garbage, wherever located, were unsightly, smelly, and hazardous to health. Sad to say, in parts of Taiwan this is still the way the system works.

Improvement is on the way, however, in that restrictions begun in 2002 greatly reduce the use of plastic bags, as well as Styrofoam and plastic food containers. Improvements in garbage collection and pick-up systems, including recycling, also began to be implemented in the 1990s and have been expanded since then. Food wastes, for example, are often separated out and sent to pig farms in the countryside. Taiwan also started using incinerators in recent years, although even that modern approach to garbage disposal has not been without its critics, especially in regard to air pollution. Regardless, Taiwan is starting to come to grips with this problem and these initiatives are addressed in Chapter 3.

Sewage

The other big contributor to water pollution is sewage. How to dispose of human and animal bodily wastes (sewage) is one of the biggest problems confronting every society. In Taiwan in the past, when the vast majority of the population lived in rural areas and the population density was not nearly as high as it is now, it was fairly easy to recycle such wastes directly back into the natural ecosystem by using them as organic fertilizers. Indeed, before the era of chemical fertilizers, farmers routinely relied on organic fertilizer, via composting, as the main or only source of field nutrients. Humans (and other animals) were in a natural, balanced ecosystem then. Population growth, urbanization, and economic development have nearly destroyed the old recycling system. Unfortunately, modern methods of dealing with sewage did not come along until recent years, well after urbanization and industrialization were under way. As the old recycling system withered, and nothing immediately replaced it, the inevitable result was rapid deterioration of Taiwan's water quality – both fresh water (rivers, lakes, underground aquifers) and coastal salt water. The population increased more than threefold between 1950 and 2004, from less than seven million to more than 22 million. The island also became a major pig producer in the post-Japanese era, with almost 11 million pigs by 1996, about one for every two people. Although the number then declined to its present roughly seven million, that many pigs still produce a great deal of waste.[30] It is not an exaggeration to describe pigs as living waste-producing factories, in that the average mature pig can generate 100–150 grams of solid waste (manure) and 33 litres of liquid waste (urine) per day, which is five to seven times the waste produced by a grown human. In addition, the water supply was further degraded by the vast amounts of pollutants emitted by the thousands of factories that arose all around the island (this topic is addressed in a later section of this chapter).

According to the TEPA, domestic sewage accounts for 44 per cent of all water pollution, pig-farm discharge for 22 per cent, and industrial wastewater for the remaining 33 per cent.[31] The pollution removal efficiencies for these in 2003 were 24 per cent, 48 per cent, and 85 per cent, respectively, for an average total of 64 per cent.[32] In other words, Taiwan appears to be doing a better job treating industrial wastewater than human and animal wastes.

Not until recent years was sewage even collected by pipe in Taipei and other urban areas. Flush toilets and piping did not begin to be introduced until well into the post-1950 KMT era. The usual practice until then for collecting human sewage was to use buckets in dwellings (the so-called 'honey buckets') that were individually emptied each morning by workers into large holding tanks on wagons or trucks that made the rounds of neighbourhoods and then hauled out to the rural areas for composting or just dumping. As piped sewer systems with flush toilets slowly began to be installed in cities, the raw sewage went directly into rivers without any treatment whatever. Moreover, the sewage commonly flowed through the urban areas in uncovered or only partially covered drainage ditches, the stench from which could be overpowering. It was a primitive and very unhealthy system ill suited to a modern urban society. There still are places in Taipei (not to mention other cities and towns) where one can smell sewage drains that have yet to be modernized. Small wonder, thus, that the rivers in the Taipei Basin became little more than huge open sewers by the 1960s and 1970s. This polluted water then reached the sea and proceeded to contaminate nearby fish and shellfish stocks. Taipei and other places suffered periodic warnings when this or that seafood (especially shellfish) was declared unsafe for human consumption (see below).The Tanshui/Keelung/Hsintien Rivers, along with the Love River in Kaohsiung, became symbols of the environmental crisis facing Taiwan.

Hazardous (toxic) waste

Industrial waste comes from factories, agricultural enterprises (especially pig farms), medical facilities (such as hospitals), and other business organizations. Medical waste totals about 90,000 tons per year, of which 13,500 tons is infectious and most of this properly handled. Agricultural waste, as already noted in the case of pig farms (above), is mostly reusable organic material and most is now disposed of in landfill sites. Our primary concern here is with pure factory waste. As of 2001, the TEPA stated that Taiwan's industrial plants produced over 18 million tons of waste, of which hazardous waste made up approximately

8 per cent (or nearly 1.5 million tons) of the total. Factory waste in total consists of three main types: metallurgy waste (32 per cent), non-metallic mineral industry waste (13 per cent), and food processing industry waste (11 per cent); the remaining 44 per cent consisted of various other forms. Hazardous waste comes mainly from three sources: electronics and electrical engineering plants produce 43 per cent, chemical plants 31 per cent, and metal industry plants 14 per cent; the remaining 12 per cent comes from other sources. In terms of regional distribution, southern Taiwan (centred on Kaohsiung) produces the greatest volume of total factory waste (52 per cent), followed by the northern region (centred on Taipei) with 28 per cent, and then central Taiwan (centred on Taichung) with 18 per cent, and eastern Taiwan (Hualien-Taitung) at just 2 per cent. Interestingly, though, the greatest volume of just hazardous waste comes from northern Taiwan (50 per cent), followed by the south (35 per cent) and central region (14 per cent).[33] About eight million tons (44 per cent) of all industrial waste is recycled, three million tons (18 per cent) are treated by waste-handling organizations, while the remaining amount, roughly seven million tons (38 per cent) is not disposed of properly.[34] This latter amount is the main problem.

Improper disposal of industrial waste is the stuff of TV dramas and Hollywood movies, but it is a very real problem in every country with a manufacturing sector. In Taiwan's case, because so many plants are small to medium size, tracking and correcting the problem is a real challenge. Industrial firms often lack sufficient knowledge about their responsibilities for handling of waste. The threat to the environment, and to humans, is not always immediately apparent to them. The owners and managers have to be educated. Moreover, smaller firms typically produce modest volumes of waste and hence do not have the economies of scale that larger firms have in terms of handling space, technology, personnel, and financial resources. Hence, it is rare for firms, of any size, to voluntarily take on recycling and proper disposal of wastes. Their bottom line is profit; proper waste handling can threaten profits. Hence, less socially responsible firms will often try to get around rules and regulations, and try to avoid unnecessary waste handling costs. One end result is illegal dumping, which has been going on in innumerable sites all over the island since Taiwan started industrializing in the 1950s. Unfortunately, such practices still occur.

Chemical wastes are supposed to be taken away from plants and properly disposed of by licensed waste handlers. However, firms often do not bother with registered waste handlers and simply hire someone to haul away their wastes. Once it leaves their premises, they tend to think it is no longer their responsibility. Throughout Taiwan there were 139 illegal toxic waste dumps known to the government in 1999, but only six legal privately owned sanitary landfills.[35] Hence, much of the industrial waste is simply dumped in local landfills every year, or wherever the dumpers can get away with it.

A variety of examples illustrate some of the dimensions of the problem:

1 In 2000, more than 33 illegal waste handlers were discovered dumping wastes all over Kaohsiung County, including toxic solvents into Kaoping

Creek, polluting the drinking water for nearby residents. Although some of the culprits in this incident were prosecuted, many observers in Taiwan think this is only the tip of the iceberg in regard to illegal dumping of hazardous wastes.[36]

2 In 2006, the TEPA caught a man who was using an incinerator on his father's farm to secretly and illegally treat hazardous waste (he was not a licensed waste handler). It was a way to make extra income, by offering low prices to unscrupulous factory owners looking for a cheap way to get rid of their wastes. Inspectors found barrels of chromium-laced toxic waste at the site. The TEPA also sadly noted that this was but the tip of the iceberg, with dozens of such operations known to exist around the island. Penalties for violating the Waste Disposal Act in this way include prison terms for up to five years and heavy fines, but apparently the low chances of getting caught encourage the illegal waste handlers to take the risk.[37]

3 In 2001 in Hsinwu Township, Taoyuan County, two local politicians were caught trying to make money as unlicensed waste handlers. They collected toxic soil for a fee, claiming to be capable of disposing of it safely, but instead dumped it on farms and a river bank in Hsinwu, polluting the farms and river, and physically threatened people who planned to report them to authorities.[38]

4 In September 2001 in the Huwei region of Yunlin County, 1,800 kg of cadmium-contaminated rice were found in warehouses in southern Taiwan. The TEPA traced the rice back to rice paddies in Huwei. Nearby chemical plants were suspected of being responsible for polluting the area's groundwater through improper disposal of wastes. Throughout the island, the TEPA found that a total of 1,024 hectares of land in 2001 were contaminated with one or more of eight kinds of heavy metals – arsenic, mercury, copper, zinc, lead, nickel, cadmium, and chromium. The Huwei incident was merely one instance of this pervasive problem, which is almost certainly greater than the statistically reported amount of land suggests.[39]

5 In the autumn of 2004, residents of Lingyun Village in Lungtan Township, Taoyuan County, complained of various ailments, including skin rashes, asthma, and headaches, and of strange smells in the groundwater. Because a recent typhoon had temporarily caused suspension of regular water supply to the area, residents were forced to rely more on groundwater. Investigation by the TEPA revealed serious pollution of the groundwater by hazardous chemicals, including phenol and methylbenzene. Investigation was under way in 2004 to find the perpetrators and make them pay to clean up the site.[40]

6 A high profile toxic dumping took place in Taoyuan County, in what one TEPA official dubbed 'Taiwan's Love Canal'. The electronics industry is normally viewed as one of the safer or 'clean' industries for the environment, but the experience in Taoyuan proves otherwise. RCA was one of the pioneer foreign companies to invest in Taiwan, back in the late 1960s when the electronics industry was just getting jump started. RCA moved produc-

tion of its black-and-white television sets from Memphis in the USA to Taoyuan, about 30 km south-west of Taipei. It was one of the first major companies to move overseas. RCA used organic solvents to clean and degrease components. Environmental standards were virtually non-existent then, so used solvents were simply poured into toilets, sewers, or thrown on the grass outside. Danger signs started to appear, such as workers becoming nauseous and dazed, even birds behaving oddly. Inspections by the Council of Labor Affairs found numerous violations of safety standards, but nothing really was done to correct the problems. Between 1970 and 1992, toxic organic solvents continued to be dumped indiscriminately on the factory grounds. The plant was purchased by General Electric in 1986 and then in 1987 sold to Thomson Multimedia (a French conglomerate). Thomson became aware of the situation and closed down the plant in 1992, and the site was sold to a local developer. By then the TEPA was in existence, and it did extensive tests on the site and found the levels of toxic solvents in the water were 960 times higher than considered safe for human use. Two of the solvents, tetrachloroethylene and trichloroethylene, are believed to be carcinogens. For more than two decades, workers and residents in the neighbourhood had been drinking and using the water, and claimed higher than normal rates of cancer. During the period 1996–8, GE and Thomson were required to clean up the site, by removing 10,000 cubic yards of toxic soil, which helped improve conditions, even though groundwater contamination remains severe. The site is still regarded as toxic and unsafe.[41]

7 In 2005, the TEPA revealed information about five abandoned factories whose locations were seriously polluted by heavy metals, mercury, pentachlorophenol, and dioxin, with severe damage to the soil and underground water in surrounding areas. The five sites are: the China Petrochemical Development Corporation's Anshun plant in Tainan (see below); the Ho Cheng Metal Industrial Corporation in Kaohsiung County; the Taiwan VCM Corporation's plant in Kaohsiung City; the CPDC's Cianjhen plant; and the Kaohsiung Ammonium Sulphate Company. The pollution included mercury concentrations 580 times greater than legal, or almost pure mercury, plus copper, zinc, arsenic, organic chlorine, and vinyl chloride. Moreover, the TEPA also revealed that there were then 175 high-risk abandoned factory sites under investigation, with another 30,000 waiting to be investigated.[42]

8 In 2004 it was revealed that residents living near the Anshun site in Tainan County used over several decades by the former state-owned Taiwan Alkali Industrial Corp (TAIC), a petrochemical firm, had dioxin levels as much as 15 times the recommended level, and high rates of cancer. The plant was built by the Japanese in 1942 to produce toxic gas for the war effort. After the war, TAIC was formed under the ROC government, to produce pentachlorophenol (PCP), used in herbicides and disinfectants, from 1964 to 1979. The plant closed down in 1979, and then in 1983 TAIC was merged with state-run China Petrochemical Development Corporation and then

some 5,000 tons of PCP were produced. The Ministry of Economic Affairs, responsible for all state-run corporations, failed to monitor pollution at the site. The dioxin appears to have got into residents' blood through fish that they obtained over the decades from a 13 ha. pool inside the polluted area. Residents were never informed of the dangers of eating fish and oysters raised there. The fish were found to have dioxin levels 10–100 times the maximum permitted by EU countries. In one family, eight people were dying from cancer. The total contaminated area was not even fully known, since some of the chemicals were thought to have spread widely through the water supply. There was concern that even higher dioxin levels could be found at other sites, in yet unreported cases of pollution.[43] The Head of the TEPA later stated that it would be impossible to clean up the polluted site, and the best that could be done is to seal off the area to prevent the pollution from spreading.[44] Victims are being compensated by the government, but the government is not taking blame for the contamination.[45]

9 In 2005, an incident occurred along the Erjen River dividing Tainan and Kaohsiung counties. This river has long been listed as one of the ten most seriously polluted rivers in Taiwan, because of sewage discharge from upstream pig and duck farms and industrial waste from plants refining and smelting metals. The TEPA has been working for several years to clean up the river and thought it was making some progress, stating that 'only' 47.2 per cent of the river remained 'seriously polluted' in 2005 (!). Then in March, the TEPA discovered 35 hidden pipes secretly discharging industrial wastewater into rivers in the area, including 21 illegal pipes going directly into the Erjen River. These pipes were discovered because a 10 km section of the Erjen River turned virtually black and inspectors were able to pinpoint the sources of the pollutants.[46] Yet one more example of how much remains to be done in educating industrialists in Taiwan and in enforcing environmental laws and regulations.

10 In 2005, the discovery of high levels of dioxin in duck eggs from Changhua County raised a storm of controversy, not only about the dioxin but also about the TEPA's handling of the crisis. Random sampling of the eggs early in the year revealed higher than acceptable dioxin levels, resulting in the culling of more than 20,000 ducks and destruction of well over a million eggs. The duck farmers were distraught, as were worried consumers, although per capita consumption of duck eggs is only 18 per year, meaning that the threat to the average consumer was actually quite low. Nonetheless, consumption of duck meat and duck eggs plummeted as panic took over and farmers and public wanted immediate knowledge of the source of contamination and punishment of the guilty party. After initial investigation, the TEPA declared the culprit to be a nearby factory that was recycling toxic ash and the factory was closed down in June. Then in September the TEPA thought the feed used on the duck farms was the culprit, but the Council of Agriculture felt the soil more likely was the problem. By December, the finger was being pointed at the Taiwan Steel Union Co. Ltd, which suppos-

edly had been dumping dioxins for some time, but the TEPA had kept quiet until the egg contamination became public. Duck farmers in the area were seeking compensation from the government, which was reluctant to do so until the full facts of the case were known. Critics felt the TEPA had not handled the situation well, delaying unnecessarily long and putting out speculation about the cause of the pollution before the true situation was clear.[47] The whole episode was a reminder of how quickly a pollution incident can get blown out of proportion, and speculation and panic set in, if the TEPA and other government agencies do not coordinate and act quickly to disseminate accurate information.

Collectively, what these few examples reveal is that the final tally on sites contaminated by toxic wastes, and the numbers of human victims and their levels of exposure, are still far from completely known.

The problem of toxins in fish has become widely publicized in Taiwan in recent years, in part because of the key role that seafood plays in the average Taiwanese diet. One study conducted by the Environmental Quality Protection Foundation showed that 47 per cent of the fish available in the marketplace in 2000 contained residues of organochlorine, one of the 12 persistent organic pollutants prohibited by the United Nations.[48] A TEPA study in 2005 found that fish from Taiwan's rivers contain high levels of polybrominated diphenyl ethers (PBDEs), which are widely used as an inexpensive flame retardant in consumer goods such as furniture, building materials, plastic products, and electronic devices. PBDEs came into use, ironically, because of consumer protection laws that demanded flame retardant properties in household products, especially plastics. The chemical is not molecularly bound to anything, and hence when the product is disposed of as waste, the PBDEs easily attach themselves to dust particles and eventually get into the food chain. The average concentration in the TEPA samples was higher than those in northern Europe, but lower than those reported in the Great Lakes of the US. The TEPA was urging people to eat fish that was either farm raised or caught in the ocean.

Mercury is another troublemaker frequently found in fish. Taiwan's people have an average of 2.4 mg of mercury per kg of hair, according to the TEPA. This compares with an average of 2.1 mg/kg in Japan, and 1 mg/kg in the US. Non-vegetarians' hair contains the highest proportion (2.54 mg/kg) of mercury, while vegetarians had only 0.32 mg/kg. Spatially, residents of Hualien had the highest rate on the island (3.09 mg/kg), while those in Taichung City the lowest (1.83 mg/kg). The correlation with seafood consumption is unmistakable (Hualien is a smaller, more traditional town on the east coast, where seafood plays an even bigger role in the diet). Moreover, those who eat deep-sea fish (tuna, swordfish, salmon) had six times the rate of those who never eat fish (3.68 mg/kg vs. 0.55 mg/kg). Fishermen in Hualien protested about the TEPA's warning to avoid deep-sea fish, claiming it caused the fish market to become stagnant, in an industry already hard pressed (see section on *Fishing* below).[49] Testing of Taipei City residents revealed concentrations as high as 7.95 mg/kg of mercury in some individuals, and 80 per cent of those tested exceeded the acceptable limit. Officials of the TEPA also contended, though, that the higher mercury content in hair was also partly the result of mercury from industrial plants getting into the soil and then into the food chain that way.[50] Interestingly, the TEPA also reported that mercury levels have actually declined slightly since 2003.[51]

It is bad enough that these kinds of incidents occur in Taiwan. Perhaps even worse is the practice of exporting toxic wastes out of Taiwan to other countries. Taiwanese companies are inventive. As environmental regulations became stricter in Taiwan in the 1990s, accompanied by rising labour costs and land prices, Taiwanese investment overseas soared, especially in China and Southeast Asia. Firms were looking for places with weak environmental laws, in addition to cheaper land and labour. One of the most notorious cases occurred in 1998, when Formosa Plastics Group (FPG), not allowed by the TEPA to bury mercury-laced mud in Taiwan, shipped 3,000 tons of the toxic mud to Cambodia, where it was dumped in a town outside of Phnom Penh, the capital.[52] FPG was found out and the company was forced to ship the waste back to Taiwan and store it in a temporary site near Kaohsiung, giving the company a tarnished reputation as well as a hefty fine from the TEPA. FPG got caught, but many other firms are not caught.[53] Just between 1993 and 1998, Taiwan shipped almost 40,000 metric tons of industrial waste overseas, much of it to China, which would seem at first glance to hardly need more waste added to the mountains of it generated within China itself.[54]

A somewhat similar incident took place in 2004, when a company in Taiwan, in collusion with a Malay company, used a fake Malaysian import permit to get approval to ship 12,000 tons of waste, which included toxic materials from dissolved waste, to Malaysia, where it was to be used to make building bricks. The conspiracy was found out, the companies heavily fined, and the waste ordered to be shipped back to Taiwan.[55] Not only are these kinds of incidents bad for the environment, but they also are not good for Taiwan's international reputation at a time of heightened tensions in the region because of the ongoing struggle between Taiwan and China over reunification. This clearly is not the way for Taiwan to win friends in the region.

Sometimes, Taiwanese companies go far away to do their polluting. Formosa Plastics Group (FPG) went all the way to Seadrift, Texas to build an enormous PVC (polyvinyl chloride) plant. FPG claimed the plant would be 'the jewel of the Texas Gulf Coast', and would emit zero toxic emissions. Instead, the EPA (USA) found massive contamination of groundwater under the plant and repeatedly fined the company. By the late 1980s the local shrimp industry was in bad shape, and residents were complaining of health problems. To be sure, FPG was not the only culprit in the area (petrochemicals are a big industry along the Gulf Coast), but it demonstrated the willingness of some Taiwanese companies to get around pollution controls in Taiwan by moving operations to wherever environmental regulations seem less strict.[56]

Another disturbing source of toxic waste is discarded batteries. About 50,000 tons of old batteries accumulate in Taiwan each year. Only just over 10 per cent of those batteries were recycled in 2002, although the goal was to be 60 per cent by 2005. An average lead-acid car battery contains just over 8 kg of lead and almost half a litre of sulphuric acid; hence, it is vital these batteries be disposed of properly. Unfortunately, some 10,000 tons of used batteries are thrown into Taiwan's incinerators or landfills each year, leaving a dangerous residue of mercury, lead, chromium, and other elements.[57] Ever eager to make money, unscrupulous entrepreneurs in Taiwan have been finding ways to make money illegally from the old batteries. Some are smuggled by ship to the mainland where they can be recharged, repacked, and sold. Some of the batteries are cut in underground factories in Taiwan, to retrieve the lead for profit (the lead from a single battery is worth several NT$ in the marketplace). The sulphuric acid, if dumped illegally, can severely pollute the water systems in Taiwan and wreak other damage.[58] The other side of the coin is the problem of below-grade batteries, manufactured in China, which are sold in Taiwan. An estimated 20,000 tons of these batteries, with average mercury content 360 times larger than those of legally controlled batteries, plus extra lead and cadmium, circulate in Taiwan each year. Because these do not meet TEPA standards, the batteries must be imported illegally, and their low price makes them attractive to bargain hunters oblivious to the environmental hazard.

In 2006 secret stockpiles of waste batteries abandoned by the military in Matsu were discovered, exposing the lack of a national system for recycling batteries. The batteries were in a depot not used for 40 years, and the military did not even know how many batteries had been dumped there. This battery incident was a reminder of the neglect of the environment during the martial law era, when the struggle with the PRC was especially intense on the outlying islands, and so many facets of government and military behaviour were hidden behind cloaks of 'national security'. The batteries on Matsu were discovered during road construction, but it was not until the local media exposed the discovery that the military started to take action; they found the burial site far larger than anticipated and were forced to call in the TEPA. Although the TEPA has been promoting battery recycling since 1999, it estimates that only 17 per cent of used batteries (not counting those used by the military) are recycled. This figure is well below the levels found in most European countries. This is one area of environmental protection where progress has been minimal.[59]

E-waste

One of the newest toxic waste problems, worldwide, is what has been dubbed 'e-waste', of which batteries (above) are but one part. E-waste refers to the disposal of discarded computers and electronic components. As information-technology (IT) industries flourish in an increasing number of countries, including Taiwan, inevitably this new source of toxic waste has emerged. Just some of the examples of toxic materials to be dealt with include: the burning of printed circuit boards at low temperatures leads to the release of extremely toxic components that can cause cancer; barium found in e-waste can damage the heart and liver; beryllium found in computer motherboards and cadmium in chip resistors and semiconductors are poisonous and could lead to cancer; chromium in floppy disks, lead in batteries and computer monitors, and mercury in alkaline batteries and fluorescent lamps also pose severe health risks. As computers and related electronic components and devices proliferate in staggering numbers, made greater in volume by built-in obsolescence and need for frequent replacement and upgrades, societies in many countries confront this problem, which makes disposal of ordinary household garbage seem tame by comparison.[60]

Comparatively speaking, Taiwan is a relatively small player, with a total e-waste in 2003 of just over 14,000 tons (compared to 2.12 million tons for the US and 1.1 million tons for Germany).[61] But, Taiwan's e-waste is generated by over 30,000 companies in the electrical and electronic equipment (EEE) industry. As with other aspects of Taiwan's environmental problems, mobilizing such a large number of firms of greatly varying sizes is a complex undertaking. Moreover, the European Union (EU), with the US and Japan likely to join in soon also, has adopted new regulations requiring manufacturers of EEE products sold in Europe to draw up a set of 'end of life' specifications for recycling their products, and to pay the bill for that recycling, if they wish to continue selling in the EU. Taiwan currently sells about US$8 billion of such products there each year, so this market is critical to the survival of Taiwan's EEE industry. These new Restrictions on Hazardous Substances (RoHS) are doubly challenging for Taiwan, because the vast majority of EEE products produced in Taiwan are not finished products but components, of which more than half are exported to China for use by Taiwanese-run companies. Thus, China is also now wrestling with these new regulations, and Taiwan will soon have to comply to satisfy import rules into China as well as the EU and other markets.[62] China was expected to have its RoHS rules in place in 2005, and Taiwan and South Korea to follow suit by 2008 at the latest.[63]

One can note the volume of e-waste in recent years within the context of many different types of so-called 'harmful' waste, as defined by the TEPA (Table 2.6). The list of waste containers, batteries, pesticide containers, abandoned motor vehicles, tyres, lubricating oil, appliances, and e-waste is a litany of the detritus of modern, industrial societies everywhere. The volume is staggering, and the problem of dealing with these items challenging.

Noise pollution

In some places noise may not be considered part of the 'environment'. Fortunately, in Taiwan it is. As anyone who has lived or stayed any length of time in Taiwan, especially in Taipei, can attest, noise is a very real part of the stresses of everyday life. For that reason, the TEPA regards noise control as a vital component of its mission. Noise cannot be seen, tasted, or smelled, but there is no question that excessive and loud noise can be detrimental to the quality of life, and to human health. Many things create noise. To provide some order to its

Table 2.6 Varieties of harmful waste, 1997–2004

Year	*Waste containers*	*Dead batteries*	*Pesticide containers*	*Abandoned motor vehicles*	*Rechargeable batteries*	*Tyres*	*Lub. oil*	*Appliances (including air conditioning)*	*E-waste*
1997	19,360	–	228	400	13,964	51,224	–	–	–
1998	126,668	14	620	187	26,285	56,630	8,008	416	139
1999	187,264	257	665	534	30,334	94,648	13,023	1,155	486
2000	225,947	632	738	504	31,688	100,283	11,996	986	947
2001	245,299	586	886	530	36,581	119,034	12,328	1,849	1,248
2002	280,959	923	961	543	32,856	103,747	9,413	1,300	1,701
2003	356,909	1,016	1,004	326	41,778	120,541	9,008	1,283	1,820
2004	351,862	1,364	1,123	416	37,739	107,191	13,325	1,285	1,930

Source: TEPA unpublished data.

Note
All figures in 1,000 units.

efforts to suppress unnecessary noise, the TEPA classifies noise in Taiwan into seven different categories:[64] (1) nearby noise; (2) folk noise (temples, weddings, funerals, etc.); (3) construction project/installation noise; (4) noisy 'facilities'; (5) vehicular noise; (6) aircraft noise; and (7) transportation noise. Each of these categories has different intensities and frequency of occurrence, and the TEPA has established maximum tolerances for them. The bottom line, however, is that noise has become progressively worse as Taiwan's population has grown, become more urban, and richer. More people, more cars and vehicles, more electronic devices, more construction – noise escalates as the pace of life quickens.

All cities tend to be noisy, of course; Taiwan's are simply noisier than some, in part because of cultural peculiarities, such as the way funerals and political campaigns are conducted. Moreover, Chinese traditionally have enjoyed '*jih-nao*', which means literally 'noise', but noise in a good way, as in 'bustling' and 'busy', the idea that it is a good thing to be surrounded by people one knows, in restaurants and other public places, and at home with all the family around. Chinese are intensely social beings, very fond of group activities, whether with family, friends, fellow workers, or students. This group consciousness, as opposed to the intense individualism of Western society, is definitely reflected in the higher noise levels one finds in Taiwan.[65]

Yet, there are limits of tolerance, even for people in Taiwan. An obvious way to measure the problem is by the number of noise complaints received by the TEPA. The volume of complaints has steadily increased since the TEPA was founded, reaching almost 30,000 a year now. In fact, noise is the leading cause of environmental complaints to the TEPA, perhaps because it is something the public feels can be remedied fairly quickly, such as by a phone call to authorities (as opposed to air and water pollution, or garbage disposal, which everyone knows are likely to take much more time to correct). In fact, however, noise is a relatively elusive culprit to control, because it is ephemeral, with no physical residue left behind, and hence proving who is at fault sometimes can be difficult. Moreover, decibel level limits can be difficult to enforce or monitor. In reality, much depends on public goodwill and willingness of the individuals or organizations to behave responsibly. The TEPA admits that 70 per cent of noise complaints cannot be verified.[66] Then there is the cultural factor noted above. One person's noise is another's enjoyment of life. A family gathering or a wedding can be a joyous occasion for the participants, but an annoyance to the neighbours. Cultural attitudes and habits are difficult to change.

'Eye' pollution

Eye pollution refers to the aesthetics of the built environment. This is the most ephemeral and subjective form of pollution and not normally considered part of the 'natural' environment or environmental problems, and hence the TEPA, for example, does not include this problem in its long list of ills that it tries to correct. The reason for this is that the built environment consists of all the elements of the landscape constructed or designed by humans, including buildings, roadways and other transportation, commercial signs, gardens and green space, and so on. It is the total urban as well as rural design. These elements of the landscape are the responsibility, of course, of many public and private organizations. Environmental bodies such as the TEPA deal with pollution of the land, air, and water, but not the design of the human landscape, even though that design may directly and indirectly influence the other forms of pollution. Eye pollution is typically more of an urban than rural problem in Taiwan, although one can just as readily find bad taste and bad design in rural villages and tourist facilities, say, as in any city. Indeed, the tourism industry is peculiarly prone to bad taste, from a Western perspective at least, reflecting the reality of cultural relativism. One culture's gem of a design is another's kitschy eyesore (China is replete with such sites).

Regardless, in highly urbanized societies such as Taiwan's, the built environment is very much a part of the complex mix of factors that collectively affect the quality of life of residents, and hence warrants at least brief mention here. Unfortunately, the built environment in Taiwan, especially in Taipei, Kaohsiung and other cities, often leaves something to be desired. Especially in years past, visitors to Taipei, for example, often remarked how ugly the city was, although the 'ugly duckling' trophy in Taiwan has long been held by Kaohsiung.[67] To be fair, the built environment has gotten better over the decades, especially in certain parts of the cities, thanks to several factors:[68] (1) the rising prosperity of Taiwan in recent years has meant more money for urban improvements, and sometimes striking and imaginative architecture in new buildings, and the design of public facilities such as the Mass Rapid Transit system and its stations; (2) the flowering of Taiwan's political and social environment since the late 1980s has helped create NGOs and civic organizations dedicated to environmental and civic improvements, accompanied by a growing environmental movement that has made people more aware of the importance of a high quality built environment; (3) public education programmes of the TEPA and other governmental bodies have also helped to foster real changes in public behaviour and expectations about the built environment, which have in turn influenced architects, designers, and urban planners; (4) the increased internationalization of Taiwan, with its globalized economy, large expatriate population, and large numbers of Taiwanese who have lived or travelled abroad, have all made Taiwanese more sophisticated and more attuned to international standards of quality and taste in built environments. There is even a 'Taiwan New Landscape Movement', ostensibly designed to fix some of the damage caused by decades of relatively unbridled industrialization, as part of the government's Challenge 2008 master plan (see also Chapter 5).[69]

Yet, it would be an exaggeration to state that Taiwan has solved problems of eye pollution. The sheer density of population and high level of urbanization in Taiwan pose extraordinary challenges to all involved in designing and building that environment. Still, Taipei today is a much handsomer, more sophisticated city than was true 20–30 years ago, and can increasingly stand up to competition with most other leading Asian cities as a desirable place to live. Kaohsiung also is a far more attractive city than it was 30 years ago, in many ways a city whose reputation has not caught up with the changes. One way to note this is found in the annual analysis of 40 of Asia's key cities that *Asiaweek* Magazine used to do each year.[70] In the 2000 analysis (the last in that series), Taipei came out tied for 4th place with Osaka, and formerly much maligned Kaohsiung ranked an impressive 12th, based on an analysis of a great variety of indices that collectively gave a quality of life score. In the critical category of 'suspended particles' (PM_{10}) Taipei ranked 7th and Kaohsiung 22nd, indicating the need for further progress in controlling air pollution, especially in Kaohsiung.[71] For more recent rankings, one can turn to the Mercer Survey, in which Taipei does not do too badly (see Box).

Ranking of cities: The Mercer QOL survey

Ranking of cities is a highly subjective activity. No matter what criteria are used, if more than one criteria is involved it ends up putting apples with oranges and trying to quantify topics that are often impossible to quantify, let alone accurately compare with one another. The final summed total of criteria (even when weighted to somehow make the criteria more comparable), from which the ranking is then made, inevitably presents a skewed, subjective comparison that should not be taken too literally. Nonetheless, rankings of all kinds of topics are routinely done, and the rankings often given undue press notice. This is definitely the case with rankings of cities, because city and national governments, and businesses, all want their individual city to look good in any regional or international comparison. Cities are like corporations, competing for market share.

One of the more widely quoted rankings of cities today is the Mercer Human Resource Consulting surveys of more than 350 cities worldwide. Mercer, an American firm, performs this task each year, and the resulting 'Quality of Living (QOL)' ranking receives much attention. Mercer's study is based on detailed evaluation of some 39 criteria, including: political, social, and economic environment; medical and health services; schools and education; public administration and transport; recreation; consumer goods; housing; natural environment; and others. Cities are ranked against New York as the base city, which has an index score of 100. The final index is mainly targeted at expatriates, to be used by them as a guide for which cities are the most comfortable to be stationed in. Hence, natives of the cities in the survey might view the criteria with different weightings and perspectives. In fact, Mercer makes clear that it is measuring the 'quality of living' not 'quality of life', which has to do with the intangibles of living in a particular city and is nearly impossible to quantify.

In the 2005 survey, almost half the top 30 scoring cities were in Western Europe, hardly surprising from a corporate perspective. Within

Asia, Singapore was the No. 1 city, but in 34th place internationally, followed by Tokyo (35th internationally), then Yokohama (37th), Kobe (40th), and Osaka (51st). Thus, Japanese cities occupied four of the five top spots among Asian cities. The top ranking city in China was Shanghai (103rd internationally). Hong Kong came out 68th internationally, while Taipei showed up in 81st place (its relatively lower ranking due to weak scores on drinking water, air pollution, leisure activities, climate and natural disasters, and traffic congestion, reflecting many of the issues discussed in this book). Kaohsiung came out in 101st position (edging out Shanghai slightly). Seoul turned up in 89th place. Honolulu came out on top among US cities. The least attractive city in the world for expatriates? Baghdad, no surprise there.

Sources: www.mercerhr.com; *The Economist*, 15 April 2006, p. 98; 'Making our cities more livable is a vital goal', *Taipei Times*, 21 May 2006, p. 8.

Regardless of rankings, the bottom line is that a person's perception or evaluation of the quality of the built environment is very much a subjective thing, influenced by many factors, including ethnicity, national origin, gender, occupation and socioeconomic class, and others. While Taiwan has come a long way and there are many handsome elements to the built environment in Taipei and elsewhere on the island, still, much remains to be done in urban renewal, developing more environmentally conscious architecture, preserving precious historical sites, providing affordable housing that is also aesthetically pleasing, maintaining cleanliness and appearance of buildings and facilities, and so on.[72]

The assault on land, flora, and fauna

Another whole dimension of the environmental challenge in Taiwan is the assault being inflicted on Taiwan's land, flora, and fauna.[73] Some of this occurs inside urban areas, of course, where these elements also exist (although there is not much natural flora or fauna remaining in Taiwan's urban areas, except perhaps rodents and insects), but the main threat is in non-urban or rural areas, including the vast mountain interior of the island, as well as the coastal waters. We have already examined portions of this assault, in the form of pollution of coastal waters, for example, or in dumping of toxic wastes that pollute the soil and eventually the fresh water system. But there are many other dimensions to the problem.

Land

Nothing is more basic than land, especially when there is not enough of it. In a high population-density island such as Taiwan, with its high level of economic development, inevitably there will be intense competition for use of land, and inevitably there will be abuse of that land. The supply of land is absolutely finite, but the demand for land relentlessly escalates.[74]

Supply of land

As already noted in Chapter 1, the main island of Taiwan has a land area of only 35,840 km^2, of which steep mountains over 1,000 metres high cover about 39 per cent of the area, while hills and terraces between 100–1,000 metres elevation make up 32 per cent, leaving the remaining 29 per cent for plains and lowlands (Figure 1.2). Thus, the island's land is divided nearly equally among the three categories of land, according to elevation and slope. That last third, though, contains most of the human landscape, including the majority of cities, towns and factories, and most of the transportation systems, but only about half the farms. By 2004, the total cultivated area had fallen to around 835,000 ha, having experienced a steady, downward slide, year by year. The government divides all cultivated land into two parts: 'paddy fields' (meaning lowland plains) and 'upland fields' (meaning cultivated land at higher elevations and on slopelands). In 2004, the former came to about 427,000 ha, or 51 per cent of total cultivated land, and the latter amounted to around 408,000 ha, or 49 per cent. Most of the upland fields are in that intermediate zone between 100–1,000 metres, but there are some areas of cultivation on slopelands at higher elevation. Anyway, cultivated land altogether comes to about 23 per cent of the island's total land area.[75]

What has happened over the years with farmland is obvious. The demands of urbanization and industrialization have put a steady squeeze on agricultural land in the plains, causing that land to be gradually converted to non-agricultural uses. In turn, more and more of the uplands and slopelands have been converted to farm fields, as the government and farmers struggle to maintain the agricultural system and food production. Even though the agricultural sector now accounts for barely 2 per cent of Taiwan's GDP, and the number of workers in the agricultural sector make up only 7.5 per cent of the total workforce, there is a strong desire by government and public to maintain as healthy an agricultural sector as possible, for the sake of food security and because of attachment to the agricultural tradition of Taiwan.[76] Most people in Taiwan may live and work in cities today, but most also take pride in their agricultural heritage and do not want to see it diminished too much, let alone eliminated completely. This attitude is common also to Japan, South Korea, and other Asian countries.

Nonetheless, there has been relentless pressure on the single most important input or requirement of agriculture – the land. The best agricultural land (paddy fields) has steadily diminished in amount. The loss came to nearly 32,000 ha in the ten-year period, 1995–2004, or an average of 3,100 ha per year.[77] Upland fields held more constant in that decade, declining from 414,000 ha to 408,000 ha. The government predicted that approximately 80,000 ha of farmland was expected to go idle or 'be used for non-agricultural purposes by 2004'.[78] In other words, land is being taken out of agriculture not only for non-agricultural purposes, but also in response to changing economic circumstances. Farming is under great pressure in Taiwan, especially since Taiwan joined the World Trade Organization (WTO) in January 2002, and must gradually lower tariffs against agricultural imports and diminish its level of subsidization of agriculture on the island[79] (see Chapter 4).

Degradation of land

Besides being taken out of cultivation entirely, land also can be degraded, or made less useful, by a number of means, including deforestation, improper cultivation, and construction – in other words, excessive human occupancy. Even in the lowland plains, erosion occurs and soil is sometimes poisoned by improper dumping of toxic materials, as already noted earlier. But it is in the higher elevations, where slopes get steeper and the erosive power of water is much greater, that the land becomes more fragile and more easily degraded. In Taiwan, the pressure for more land for agriculture, transportation, and recreation has resulted in some real problems with degradation in the slopelands. Excessive or improper use of those lands has come into conflict with nature, made worse by natural disasters that are taking a greater toll than would otherwise be the case without such a large human presence.

Slopeland cultivation

Take the matter of slopeland cultivation and its consequences. Slopelands are defined by the government as having either an elevation of more than 100 metres or a gradient of over 5 per cent. Taiwan farmers have long used slopelands for cultivation, or to exploit natural forest for commercially usable timber, such as the camphor tree (see Chapter 1). But the impact was relatively limited through the Japanese era and into the early years of the ROC period from 1950 onward. As the agricultural sector became increasingly stressed, from the 1970s on, however, farmers began to move more and more into slopelands, and at higher elevations and onto steeper slopes. Most of this cultivation was illegal, on public land not owned by the farmers. It was not uncommon in the 1970s and 1980s to see very steep slopelands, with no terracing whatever, being used to grow crops of various kinds, but especially high value crops including tree fruits (citrus, mango, papaya, guava, apples, and others), tea, vegetables, and betel nuts. Amazingly, the crops sometimes were planted in rows running up and down the hillsides, rather than along the contour, maximizing run-off and displaying almost total disregard for soil conservation practices. The government, after the fact, attempted to bring some order to this process, with a slopeland development programme that was controversial and only partially attacked the problem. The government would attempt to impose terracing and more ecologically sound cultivation practices to existing illegal slopeland farming. The government also opened up new slopelands with development schemes, laying out many thousands of hectares of new land geared especially for upland cultivation, as the government tried to help farmers maintain their livelihoods. The result of the slopeland development programme was to greatly boost production of certain cash crops in Taiwan, enhancing the quality of the food supply in the urban areas and in some cases providing Taiwan with some lucrative export products. But the downside was to increase the risk of soil erosion and landslides. As noted above, roughly 49 per cent (over 400,000 ha) of all cultivated land is now officially in this slopeland zone, although not all of this land is equally at risk, of course. The Council of Agriculture classified 32,000 ha of slopeland as 'agriculturally overdeveloped' in 2000, a somewhat ambiguous term, and almost certainly an underestimate.[80]

The betel nut

One of the biggest problem crops in the slopelands has been the betel nut palm. The betel nut is a very old commodity in Chinese agriculture, originally used in traditional Chinese herbal medicine to treat intestinal disorders and parasitic infections. The betel nut in modern times, however, is primarily used as a mild narcotic, sometimes called 'Taiwan chewing gum', because the pulpy nut, usually flavoured with a lime paste, is chewed like a stick of gum and stimulates blood circulation, and is favoured by labourers as a means to increase stamina and productivity. In a manner of speaking, the betel nut is the Taiwanese equivalent of chewing tobacco in the US. It is addictive, and if chewed often enough can cause oral cancer and other medical problems. The risks, however, do not seem to deter demand or consumption. As Taiwan urbanized and industrialized, it was commonly thought that demand for betel nuts would fade away and be mostly a memory of a long-gone rural Taiwan. Instead, the betel nut has become, along with cigarette smoking, a mainstay particularly of the blue collar class and physical labourers, usually less educated and less affluent members of Taiwanese society.[81] The percentage of betel nut chewers among Taiwan's 16.7 million adults has stayed around 9 per cent in recent years (or more than 1.5 million) and some two million people rely on the business of growing or marketing the nuts for some part of their livelihood. The nuts are typically sold at roadside stands, or by car side vendors at stop lights, usually by young girls in skimpy costumes as a sales enticement (nearly all betel nut chewers are male).[82] In short, the betel nut has become a significant component of the agricultural sector, and persists as a remarkably durable element of traditional Taiwanese culture.

The impact on slopelands, unfortunately, is not good. The planted area of betel nut palms in 2004 came to over 52,000 ha, more than 6 per cent of all the cultivated land in Taiwan, with about one-sixth of the planted area on slopelands.[83] Betel nuts are Taiwan's second largest crop after rice in production volume, with a value of around US$350 million a year. This is hardly surprising, because no other crop is as profitable. The investment is small, and after five years the trees start their production cycle of twice-yearly harvests for up to 40 years. A single hectare in 2003 could bring in about US$120 income.[84] The problem with the betel nut palm tree is that it has few leaves and shallow roots. The tree is unmistakable in appearance and stands out strikingly in the landscape. On the slopelands, the trees are less able to retain moisture and provide shade than natural forest (or even fruit trees). Hence, they destabilize the soil, which becomes more vulnerable to erosion in times of heavy rainfall. Government policy has shifted in recent years away from the previous approach, which was basically to neither support nor actively forbid such cultivation, to the current approach, which is to provide subsidies to betel nut farmers to get them to turn their plots back into forest. The government is moving cautiously on this front, however, because the total industry accounts for nearly US$3 billion in Taiwan's economy. Economic alternatives have to be found for the people involved, and betel nut chewers have to be somehow weaned off of their addiction. No easy task, especially when cigarette smoking is being discouraged as well.[85] With so many other problems, many of them far more serious than betel nuts, weighing on the government, it is unlikely that this uniquely Taiwanese cultural habit will be eliminated any time soon.

Taipei Basin

Another impact on the slopelands comes from construction, especially housing and roads. One of the most notorious sites is the Taipei Basin, an area of repeated troubles because of the huge population, excessive road and bridge construction, as well as housing built on steep hillsides that surround the basin. The Taipei Basin has really been a disaster long waiting to happen.[86] Over the last 50 years the population within the basin increased dramatically, from a few hundred thousand at the end of the Japanese era to the present 5–6 million crammed inside what used to be a lake until about 300 years ago. As the former lakebed flatlands became filled, housing demand pushed construction of apartments, houses, and other structures up onto the surrounding slopelands. Much of this construction, especially in the earlier decades, was largely unregulated or not well engineered and designed. In spite of the Slopeland Development & Construction Management Measures Law passed in 1983, regulation of new construction was still lax and permits already issued were exempt from the new regulations anyway. Landslides became increasingly common. One of the most notorious instances took place in 1997, when the Lincoln Mansions in Hsichih in Taipei County, completed in 1993 under pre-1983 regulations, partially toppled in a typhoon, killing 28 residents. Nearly 1,000 households were squeezed into just 5 ha at this housing project.[87]

Hsichih has been one of the most disaster-prone areas within the Taipei Basin, because of its location at a particularly vulnerable stretch of the Keelung River, one of the three tributaries that flow through the basin. At times of heavy rains, often associated with typhoons, flooding frequently occurs in Hsichih (and elsewhere in the basin). Typhoons Zeb and Babs in 1998 inundated some 6,000 households with water up to five metres deep. Typhoon Xangsane flooded out 10,000 households to depths of eight metres in 2000. Then in 2001, Typhoon Nari flooded out almost all the 160,000 people in Hsichih and left 51 dead in the basin. The relatively new Taipei Mass Rapid Transit system also had sections flooded out, causing massive transportation snarls and huge financial losses.[88] In 2004, Typhoon Aere wreaked havoc across much of northern Taiwan, including Taipei, causing major flooding again in the basin, this time especially serious in Sanchung City and the Tanshui River, and again flooding parts of the MRT system.[89] The problem of flooding in the basin is partly caused by slopeland construction, and just too much paving over of the ground along the Keelung and other rivers in the basin, all of which increase run-off. More than 60 bridges span the Keelung alone. The river runs through a highly concreted urban environment.

Moreover, before the Feitsui Reservoir was completed in 1987 to provide water for Taipei, groundwater was being pumped excessively from under the Taipei basin and land subsidence became a serious problem, leaving the basin further less resistant to flooding.[90] In short, the Taipei Basin is a prime example of inadequate or incorrect policies to protect from natural calamities. Now, various government authorities are embarked on a massive flood prevention programme in the Basin, which hopefully will ameliorate the impact of future typhoons and rains (see Chapter 3). Beyond the Taipei Basin, however, land subsidence has been a widespread problem along much of Taiwan's coastline, especially the south-west coast, as a result of excessive pumping of underground aquifers to supply industrial zones and aquaculture that have flourished in that region in recent decades. As noted earlier, including in Chapter 1, that is an area of relatively lower rainfall, and water shortages are increasingly common. Hence, human pressure on underground aquifers is intense.

These disasters could not be totally prevented, of course. But what could have been lessened were the flooding and loss of property and human lives, and the severe disruption to the economy that results.

Flora

At the dawn of the historical era, when Chinese settlers first began to arrive on Taiwan, the island was almost entirely covered in lush forests of various kinds. Then, the natural vegetation in the plains was gradually removed to make way for paddy fields and farms and other human land uses; thus, the forest retreated more and more to the uplands or slopelands, especially the mountain interior above 1,000 metre elevation. During the Qing era (1683–1895), the mountain interior was almost entirely untouched, with the aborigines living mostly unmolested in their mountain fastness. Toward the close of the Qing era, imperial officials started to develop an interest in Taiwan's forests and set up a bureau to begin harvesting the island's trees, but the takeover by Japan in 1895 cut short their belated efforts.[91] In the Japanese era (1895–1945), there was an effort to begin to exploit some of the forest, especially around the perimeter of the interior, but most of the mountain interior remained out of Japanese control and hence untouched except for shifting cultivation by the aborigines.[92] It was only after the ROC retreated to Taiwan in 1949 that a concerted effort began to open up the interior, as part of the process of developing the whole island. Thus, the first really comprehensive scientific survey of the island's forest resources took place in the early 1950s, which revealed a still-lush landscape of many forest types, because of the sharp elevation changes on the island. The forested area at that time totalled just over two million ha, or about 20,200 km^2, roughly 56 per cent of the island's land area. Hardwoods covered 72 per cent, conifers 19 per cent, bamboo 6 per cent, and mixed conifers/hardwoods the remaining 3 per cent. Conifers predominated in higher elevations, hardwoods in lower elevations.[93] Today, the total forest cover remains basically unchanged in area, at 2.1 million ha, but with significant changes in the mix of forest types, with conifers about the same at 21 per cent, but hardwoods decreased to 53 per cent, mixed hardwoods/conifers much greater in area at 19 per cent, and bamboo up at 7 per cent. National forest accounts for 77 per cent of total forest land.[94] These changes are the result of logging operations, deforestation for slopeland cultivation, and reforestation efforts over the decades.

Government policies toward the forest cover in the early decades of KMT rule were not benign. Driven by a fierce determination to build up the economic strength of Taiwan in the face of the Communist threat from the mainland, the ROC government saw the forests primarily as a resource to be exploited. A government-run logging industry was established, through the Taiwan Forest Administration, along with a private sector operation. The government focused on high-grade conifers in the national forests which covered three-quarters of the forest area, while private operators cut anything marketable in the remaining one-quarter. Six government-run sawmills handled most of the milling of timber. One of the most important of these was the Ali Shan (Mount Ali) logging station, begun by the Japanese in 1913, in Chiayi County. A 64 km-long narrow-gauge railway, a marvel of railway engineering at the time, was built by the Japanese to get the logs down the mountainside. Wood, in the form of logs and sawn lumber, including plywood, was thus an important export product in the 1950s and 1960s. Even the age-old harvesting of camphor trees continued into the 1960s until the worldwide demand for natural camphor withered away.[95]

By the 1980s, as environmental consciousness increased in Taiwan, the government slowly began to rethink its position on forest resources. In 1991 it became illegal to cut trees in natural forests for commercial purposes. In 1999 the Forestry Bureau finally took over responsibility for managing the island's forests, putting the resource in the hands of the central government for the first time. The focus now is very much on reforestation and doing everything possible to prevent future landslides. For example, the Bureau has completed the construction of an 'Ecological Corridor of the Central Mountain Ridge', which stretches 300 km north–south and covers 630,000 ha of land, connecting three major wildlife habitats (Chilan, Danda, Chuanshan) with three national parks (Hsuehpa, Taroko, and Yushan). The corridor is supposed to be sustainable, free of unsuitable land use, and effective in preserving biodiversity.[96]

Fauna

When population density is high, and humans are competing for the same space as wildlife, the winner is rarely the wildlife. Such is Taiwan. Moreover, the Chinese in their long history are not renowned for being kind to wildlife. Throughout most of the history of the mainland, as well as Taiwan, wildlife was viewed as a threat or danger, source of food, or source of ingredients for traditional Chinese medicine, sometimes all three in the same animal. Wildlife was often a source of income, too, especially in remote rural areas where economic opportunity was limited. To this day, one can find numerous locations in rural China where wildlife is still hunted or trapped and then sold or used for the above purposes. Restaurants, especially outside of cities, commonly have cages of live creatures stacked up for patrons to pick and choose their selection for the meal. Unfortunately, this was quite common also in Taiwan until just recent times. There are still remote areas of the island where some people, especially aborigines, still dine on wildlife. The snake, especially, has persisted as one of the enduring creatures in traditional food and culture on the island.[97] The idea of protecting and appreciating wildlife for its own sake was a totally foreign concept that did not begin to gain acceptance until the environmental movement got under way in the 1980s and after.

Attitudes among some of the people, unfortunately, have yet to change. The latest demonstration of this is the practice of catching wild birds, or importing them from abroad, for sale to local religious groups, which release the birds as part of their ritual (supposedly to gain 'redemption'). Some 500 religious groups are involved. The problem is that most of the birds (an estimated 90 per cent) die before being released, or soon after release, because they are unable to survive in the wild. An estimated 600,000 birds are released each year. Hence, the total volume of birds being processed is in the millions. The volume is so great, in fact, that 60 per cent of the hundreds of pet shops around the island have been found to stock birds for this activity, which has become a big and profitable business. Religion and business have become entwined on this issue, and now conservationists and the government are trying to fix the problem, by amending the Animal Protection Law to prohibit the practice of releasing animals into the wild.[98] However, making the activity illegal, and imposing stiff fines, are not likely to stamp out the practice, especially the catching of domestic wild birds. Demand for the birds has to be targeted. Those believing in the religious benefits of this practice must be educated otherwise, and that will take time and effort.

Another demonstration of a persistent problem came to light in 2006 with a report about widespread illegal hunting in the Dawushan Nature Reserve in Taitung County, which was said to have become a 'hunter's paradise'. Even worse, the hunters were not even shooting animals for food, but simply doing it for the pleasure of shooting, and then leaving the carcasses where they fell. The evidence was photographed and revealed publicly by environmentalists, who also claimed that the police were in collusion with the hunters.[99] In another much-publicized instance, the Forestry Bureau was forced to increase restrictions on people travelling on the Danda (Tanta) Forest Trail, which traverses the Central Mountains between Nantou and Hualien counties, because of widespread poaching. The trail is maintained by Taiwan Power for its power lines in the area. Unfortunately, an unintended consequence has been to make it that much easier for poachers to obtain wild animal meat and wild plants for sale to restaurants near the trail. Some animals reportedly fetch high prices, making for a lucrative but illegal business. If caught, poachers are liable for up to five years in prison and fines up to NT$1.5 million (about US$45,000), but they are rarely caught or prosecuted. The issue is muddied by the fact that the government allows aborigines to hunt within limits to maintain their 'traditional lifestyle', but some aborigines do not stop at their limit.[100] One can only guess how common this is in other parts of Taiwan's wilderness areas.

Because of Taiwan's subtropical location, varying elevation zones, and with more than half the area still in forest, the island has a remarkably varied, but highly vulnerable, fauna. Still today, there are 61 species of mammals, more than 400 species of birds, 92 species of reptiles, 30 species of amphibians, 140 species of freshwater fish, and an estimated 50,000 species of insects, including more than 400 species of butterfly. Snakes of many types, a number of them quite poisonous, still abound. A few species of animals, though, are so rare now they are regarded as national treasures and are protected as endangered species. Currently, 174 native animals are on the list, including the Formosan black bear, the Formosan Sika deer, and the Formosan landlocked salmon. See below for examples of these endangered creatures and their habitats:[101]

1 The low mountains near the Taipei Basin are home to the emerald tree frog and green tree frog, as well as other amphibians.
2 On the western flank of the Central Mountain Range, in the Tahsueh Shan, the area above 2,000 metres is home to the Formosan black bear, the largest land mammal on the island.
3 Around the latitude of the Tropic of Cancer, on the slopes of Yu Shan (Jade Mountain), Taiwan's highest mountain, are found a number of important creatures, including the Formosan macaque, wild boar, sambar deer, muntjac deer, the Chinese pangolin, Mikado pheasant, Swinhoe's pheasant, and many others.
4 In the southern part of the island, in the Central Range, is the last refuge of the Formosan clouded leopard.

Of course, most people living in Taiwan never personally see these rare creatures, even if the people do venture into the animals' native habitats. At best, most people might see these creatures in the Taipei Zoo. But to an increasing number of people, it is sufficient for them just to know that these creatures still survive on Taiwan in their natural state.

The Formosan landlocked salmon is an example of a species hovering on the brink of extinction that appears to be making a recovery. Regarded as a 'glacial relic', this salmon species requires cool, clean water to survive, and was once abundant in the upper Tachia River in central Taiwan. Over-development of that region led to a drastic decline in the salmon's numbers and it appeared doomed. Now, the fish survives within the boundaries of the Sheipa National Park.

Another example is the endangered black-faced spoonbill, which visits the island every year in its migratory movements, stopping normally at the Chiku wetland in Tainan County in southern Taiwan. Taiwan officially has 16 areas on the island listed as important wetlands, including Chiku. Unfortunately, these wetlands have been encroached upon and damaged by urban/industrial growth, draining of low-lying areas, coastal pollution of water, excavation of sand and gravel from stream beds, and other sorts of human activities. These low-lying wetlands are not only habitats for many species of birds and other creatures, but also play an important role in the natural ecology of the island, including serving as filters of run-off and to minimize flooding. Hence, protecting the wetlands has benefits extending well beyond just preserving endangered species.

Unfortunately, sometimes the behaviour of a species is not conducive to a good relationship with humans. An example is the protected Formosan macaque (a kind of monkey). Farmers in central Taiwan recently protested for relief from the government because the monkeys were taking their fruit, ravaging their farmland, and sometimes even harassing women and children.[102] With no natural enemies, and knowing humans will not hurt them, the macaques apparently have become quite brazen. Recently, macaques living in Kaohsiung's Shou Shan natural conservation area were found to be infected with a virus that could seriously infect humans. People were being advised to avoid feeding or having contact with the macaques in that area as a result.[103] The macaque thus is an example of the unintended side effects that can result from putting a species on the endangered list.[104]

Taiwan also sometimes suffers the consequences of invasive species, a worldwide problem, most notably in such places as Australia, Hawaii, Guam, and others. In Taiwan's case, the most recent problem has been the red fire ant, which is believed to have entered Taiwan on cargo ships from the US, and has been found in 20 cities and rural townships around the island.[105] The problem is greatest in the counties of Taoyuan, Taipei, and Chiayi, and regarded with enough seriousness by the government that it recently established a National Red Fire Ant Control Centre to disseminate information about control measures to farmers and others.[106] Officials talked optimistically about being able to get rid of the pest within three years, but the experience of other countries in trying to control this and other invasive species does not suggest that such optimism is warranted. Invasive plants are also a problem. Most recently, the government has been worried about the potential danger from a parasitic weed known as *dodder*. The dodder is a leafless parasitical vine that attaches itself to another plant and is nourished by it, threatening the life of the host plant. Dodders are now common in not only coastal areas but also urban areas in Taiwan.[107]

The struggle to preserve Taiwan's wildlife, anyway, is still in its infancy. For one thing, it is difficult to root out the use of animal parts in traditional Chinese medicine. Cultural traditions die hard. Even though importation or use of certain items, such as tiger parts or rhino horn, has been outlawed since the 1980s, a black market still exists. Enforcement is difficult. The desire or need to maintain livelihoods also plays a critical role.

Fishing

This is apparent in the case of fishing, both domestic and international, in which Taiwan participates. Fish and seafood products have long played a key role in the diet of the Taiwanese. The native cuisine incorporates seafood products of an amazing variety. A large percentage of the protein in the diet is derived from such. The shallow coastal waters on the western side, and prevailing ocean currents, result in good conditions for fisheries to thrive. Traditionally, many thousands of native Taiwanese engaged in fisheries since the initial settlement of the island. Numerous fishing villages existed along the coast, especially the west (the east coast has relatively few natural harbours, because of the steep mountains and deep ocean trench just off the coast). Today, one can still find a number of these old fishing ports with their distinctively styled and colourfully decorated boats docked there, reminders of Taiwan's maritime heritage. Before the ROC era, however, most fishing was limited to near coastal waters. Even during the Japanese era, there was not much development of deep-sea fishing far from Taiwan.[108]

Taiwan's total fish production grew from an output of 120,000 tons in 1952 to 1.4 million tons in 2002, with production value doubling in that period to over US$3 billion. This made Taiwan the 20th largest fishing nation in the world, and one of the six major deep-sea fishing nations. Taiwan's tuna and squid fisheries ranked 2nd and 3rd in the world, respectively.[109]

In recent decades the fishing industry has become quite modernized and globalized, with ever larger deep-sea boats, and very aggressive in the face of intense international competition and declining deep-sea fish stocks. A good example is the case of the tuna-fishing industry.[110] Many Asian (and other) nations now compete for the Pacific's dwindling supply of tuna. Around two million tons of Pacific tuna worth up to US$2 billion are caught every year. Most of these tuna are caught by purse seiner vessels, which round up the tuna in a large net. As technology advances, newer 'super' seiners are able to take up to 11,000 tons of tuna a year (twice the amount of older boats); Taiwan is reported to have developed new seiners with 20,000 ton capabilities; some of Taiwan's vessels fly flags of convenience from the Marshall Islands and Vanuatu, in an effort to skirt international regulations. Other nations are crying foul, as a result. The UN is trying to bring order to the competition, through a new commission headquartered in the Federated States of Micronesia. The hope is that Taiwan will play a constructive role in this endeavour. The UN estimates that more than 70 per cent of the world's commercially important fish stocks are either over-exploited, depleted, slowly recovering, or close to the maximum sustainable level of exploitation.

Taiwan is a member, along with Japan, Australia, New Zealand, and South Korea, of the Commission for Conservation of Southern Bluefin Tuna, one of the endangered species. The worldwide catch is limited to about 14,000 tons. Most of the bluefin catch each year ends up in Japan (total tuna consumption there from all varieties was 150,000 tons in 2005). Japan had a quota in 2005 of 6,065 tons, but exceeded that by 25 per cent and got a lot of criticism, as well as a reduction in its 2006 quota to compensate.[111] The tuna business is but one example of numerous international environmental problems or issues in which Taiwan is involved, and Taiwan has much to gain or lose by how it responds to these international initiatives. The FAO reports that the global tuna haul (of all varieties combined) increased from just 200,000 tons in the early 1950s to a peak of 4.3 million tons in 2003. Since then, catches have been falling and prices rising, with the mature blue-fin tuna the most prized (especially by the Japanese, for sashimi and sushi).[112]

Taiwan also participates in the International Commission for the Conservation of Atlantic Tunas (ICCAT). Unfortunately, Taiwan's allowable catch quota for Atlantic big-eye tuna was slashed by 70 per cent in late 2005 because of Taiwan's past over-fishing, illegal fishing under a flag of convenience, and 'fish laundering' (incorrect reporting of catch). This was the most severe punitive action taken in the history of international fishing, an action that could have cost Taiwan as much as US$300 million a year (90 per cent of Taiwan's catch is sold to Japan). Thus, Taiwan began taking steps to control its sometimes over-zealous fishing boat crews. In 2006, for the first time, Taiwan sent a patrol vessel into the Atlantic to monitor its fishing boats to prevent further over-fishing and other irregularities. The ICCAT had cut Taiwan's allowable big-eye tuna fishing quota from 14,900 tons (2005) to just 4,600 tons in 2006,[113] but in late 2006 the ICCAT overturned its previous decision and restored Taiwan's quota to 14,900 tons for 2007. This was in response to what it saw as positive reforms undertaken by Taiwan, including reducing the size of the fleet. Previously, Taiwan had a fleet of 614 vessels (42 per cent of global large-scale tuna long-line vessels). Now, a maximum of 64 vessels would work in the Atlantic, and in 2008 would drop to 60. In addition, the vessels would be required to report their daily fishing volume via monitoring systems or radio, and the Taiwan Fisheries Agency would submit reports every six months to ICCAT and also stop Taiwanese vessels from engaging in any further illegal or unregulated activities.[114]

Observers have argued that what Taiwan needs is a new Ministry of Ocean Affairs in the government to oversee all maritime affairs, including the fishing industry, and to monitor Taiwan's compliance with international fishing agreements, as well as to help maintain all facets of the fishing industry and community in Taiwan. The issue has political and strategic overtones as well, in the eyes of critics, in that Taiwan is tangled up in the ongoing dispute over the Tiaoyutai Islands just north-east of Taiwan that are claimed by Japan, the PRC, and Taiwan. There have been clashes in the area between fleets from all three parties seeking to demonstrate their side's claim to the islands.

The other big trend in fisheries is what is known as 'aquaculture', or the raising of fish, shrimp, eels, and other sea products in ponds along the coast.

Taiwan eels, for example, are prized in Japan, and account for about 8 per cent of all fish production value. Aquaculture has become big business in Taiwan in recent decades, often through conversion of rice paddies to fish ponds. Sometimes this has also involved illegal construction of ponds in the shallow shore waters along the coast, causing damage to the natural ecosystem. The main production now comes from inland ponds using pumped groundwater mixed with sea water, which causes land subsidence. The rate of subsidence has been rising most rapidly in Yunlin and Chiayi counties along the south-west coast. In 2003 the subsidence reached over 12 cm in Yunlin and nearly 9 cm in Chiayi.[115] This land subsidence is caused not just by pond aquaculture, of course, but also by other water users, including industries that have sprung up in coastal development projects (see Chapters 1 and 4). Although a shrimp disease since the late 1980s has caused a decrease in these ponds, they still exist because the government, in effect, allows fishers to use this land along the shore as a way to help boost their incomes.

A good way to see the importance of aquaculture today is in terms of production value. In 2004, Taiwan had a total production of over US$3 billion in fisheries (Table 2.7). The government divides that production into six classes. The largest was 'far-sea fisheries' (including the tuna industry discussed above) at nearly 48 per cent of total production value, followed by 'inland aquaculture' (fish ponds, etc.) at 29 per cent, then 'offshore fisheries' at almost 14 per cent, then 'coastal fisheries' with barely 7 per cent, and 'marine aquaculture' (raising of shell fish, seaweeds, and other products) at about 3 per cent, with 'inland water fisheries' at a fraction of 1 per cent. Commercial fishing in the streams and rivers of Taiwan obviously is negligible, due to a combination of the shallowness of such waters and the problems of river pollution. By contrast, deep-sea saltwater fishing (both far-sea and offshore) is very important to Taiwan, accounting collectively for two-thirds of total fisheries production. How long that can last in the face of threats to such key parts as tuna fishing remains to be seen. Tuna fishing alone accounts for over a third of all fisheries production value, or thus half of all deep-sea fisheries revenues. Aquaculture, both inland and marine, account for 30 per cent of all fisheries production, which means this

Table 2.7 Taiwan's fisheries production/population, 2004

Type of fishing	*Production value (US$m)*	*% of total*	*Population*	*% of total*
Far-sea fisheries	1,468	47.6	19,527	4.27
Offshore fisheries	419	13.6	105,562	23.09
Coastal fisheries	211	6.8	184,037	40.27
Inland aquaculture	882	28.6	117,602	25.73
Marine aquaculture	104	3.4	23,590	5.16
Inland water fisheries	<1	–	6,722	1.47
Totals	3,084	100	457,040	100

Source: *Agricultural Statistics Yearbook 2004*, pp. 138–9, 200–1.

kind of business is very important in certain parts of Taiwan, especially along the south-west coast where it is concentrated. How fishers there respond to the environmental/ecological problems associated with the business is as yet unclear. One response that has emerged is for fishers to try to supplement their income by engaging in ecotourism in the summer, by running tours by raft or boat for watching whales and dolphins in offshore waters.

One other gauge of fisheries' still relatively strong importance to the agricultural sector and economy of Taiwan is in employment in the industry. The number of fishers' households actually bottomed out at about 113,000 in 1993, but has slowly increased since then, to 133,000 in 2004.[116] Yet, families are shrinking in size in Taiwan, so the population of fishers' households has steadily dwindled, reaching about 457,000 people in 2004. Still, that was about 12 per cent of the total farm population of Taiwan. Of that total, offshore and coastal fisheries accounted for 63 per cent of the fisheries population, while inland aquaculture accounted for another 26 per cent. The data in Table 2.7 also seem to indicate that coastal fisheries still occupy a sizeable number of people but are not very lucrative, indicative perhaps of pollution's impact on coastal waters. By contrast, far-sea fisheries occupy a small number of people for much bigger profits.

In short, while fisheries are a relatively minor part of the economy today, a significant number of people still depend on them for their livelihood, and of course a far larger number of people in Taiwan depend on the output of this industry for an important part of their diet. Hence, the relationship between fisheries of all kinds and environmental problems is extremely important.

Global environmental issues

Finally, in the long list of environmental problems of today, the people of Taiwan must also deal with environmental issues that go far beyond the confines of their small island. These are the global issues that have come to the forefront of world concern in recent decades, largely as a result of the globe's exploding population and rapidly increasing consumption of fossil fuels. Taiwan may be but a bit player on this global stage, compared with the giants such as the US, China, India, and Russia. Nonetheless, Taiwan, with its highly industrialized and energy-hungry society, is a player and has to confront its role in the environmental problems that have arisen. The problems that result are global warming, ozone layer depletion, acid rain, and the risks from nuclear energy.

Global warming

Global warming has been brought inescapably to the public's attention in recent times, sometimes through the media and sometimes through simple observation of weather behaviour.[117] The problem of global warming results from the fact that the burning of fossil fuels releases various so-called 'greenhouse gases' (GHG) into the atmosphere. The most important of these is carbon dioxide

(CO_2), but ozone (O_3) and oxides of nitrogen and sulphur are also involved. As fossil fuels are burned, CO_2 collects as an atmospheric layer that reflects the sun's heat back toward Earth, heating the planet in a process known as 'global warming'.[118] Scientists more or less agree on the process, but there are wide differences of opinion as to the severity of the problem, how fast it occurs, who is most to blame, and how to fix it. Scientists point to some disturbing facts. Over the past century, carbon dioxide concentrations in the earth's atmosphere have risen by a third, with 85 per cent of the world's energy derived from fossil fuels, and the scientists argue that unless emissions are slashed by some two-thirds worldwide, the CO_2 concentrations are projected to rise to double that of pre-Industrial Revolution levels. The world's atmosphere now contains about 380 parts per million of carbon dioxide, compared with 280 parts per million in 1800.[119] Thus the new level would be 560 parts per million around 2050. Some scientists believe the Earth can tolerate CO_2 levels of about 550 parts per million before dangerous climate effects (extreme weather, regional droughts, significant rises in sea level, etc.) would be felt.[120] This worries many, as the warning signs of global warming are already beginning to be seen. For example: major ice formations around the globe are rapidly melting and retreating, such as the glaciers of Alaska, the permanent ice mass on top of Mt. Kilimanjaro in Kenya, ice fields in the Andes of Bolivia and in the Himalayas, the Antarctic and Arctic ice packs; the climate along the Arctic coast of Alaska is already noticeably warmer, as it is in other parts of the world.[121] Significant melting of the ice caps is especially worrisome, as it could lead to major rises in sea level that would lead to flooding of many of the world's coastal regions with their numerous cities and large populations. The potential is there for disasters seen so far only in science fiction movies, such as *The Day After Tomorrow* (2003). A much more realistic depiction of the danger was dramatically presented in Al Gore's highly acclaimed documentary, *An Inconvenient Truth* (2006).

Where does Taiwan stand in global emissions of carbon dioxide? In 1978, Taiwan's total energy supply was 28.8 million kilolitres of oil equivalent (KLOE), of which domestic energy production accounted for 19 per cent of the total. By 1998, 20 years later, total energy supply had soared to 92.3 million KLOE, of which domestic production had dropped to less than 4 per cent. In short, Taiwan was not only increasingly dependent on imported energy (primarily coal and oil), but also thus generating ever greater amounts of GHG. In 1990, Taiwan's total GHG emissions came to 149 million tons in CO_2 equivalent, of which 83 per cent was CO_2, 15 per cent was methane, and 2 per cent was NO_2. The TEPA notes that the energy sector produced the largest amount of GHG, totalling about 113 million tons in 1990, ranking Taiwan 27th compared with 159 other countries, or about 0.5 per cent of the global total. The emission per capita was about 5.44 tons (35th out of 126 countries). By comparison, the emission per capita of OECD countries in 1990 was 12.14 tons. In 1997, CO_2 emissions from the energy sector totalled about 180 million tons (22nd out of 135 countries), or about 0.8 per cent of the global total. The emission per capita was about 8.11 tons, ranked 31st.[122] According to the TEPA, Taiwan's carbon dioxide emissions in the energy sector increased to 255 million tons in 2003, compared with the 113 million tons noted above in 1990, in other words a more than a two-fold increase in somewhat over a decade. This means that Taiwan produces more CO_2 than all but 21 nations, making Taiwan a disproportionate global polluter in relation to its population.[123] However one compares the data, the bottom line is that Taiwan is rising in the ranks of global air polluters. If Taiwan were to do nothing to stop this upward trend, the TEPA estimates the island's total carbon dioxide emissions would reach 501 million tons by 2020, with a per capita emission of 19.6 tons. Even though Taiwan is not recognized as a sovereign nation by most countries of the world, the government feels it is important for Taiwan to play a responsible role and do its share of reducing greenhouse gases and helping in the fight against global warming (see Chapter 3).

Ozone layer depletion

By the early 1970s, scientists came to the conclusion that the then-widespread use of chlorofluorocarbons (CFCs) is destructive to stratospheric ozone, which serves as a protective layer that screens out dangerous ultraviolet radiation from reaching the surface of the planet. The ozone layer was beginning to be depleted, because of chlorine molecules elevated into the stratosphere by industrial compounds such as CFCs. This theory was verified in 1985 with the discovery of an ozone 'hole' over the Antarctic Circle.[124] CFCs were being used primarily in refrigerants, solvents, foam blowing, and propellants of aerosols. Taiwan used them extensively as it industrialized and developed in the post-1950 era, with consumption peaking at 16,255 tons in 1988. Then Taiwan started imposing limitations on use of CFCs, and other known ozone depleting substances, such as halons and methyl bromide. Legal consumption was reduced to zero by 1996, although illegal smuggling of these substances persisted through the 1990s. Taiwan monitors the thickness of the ozone layer above the island on a regular basis, and so far there has not been a problem insofar as Taiwan is concerned. But Taiwan has tried to be cooperative in international efforts, even though not a signatory to the Montreal Protocol of 1987, to help bring the problem of ozone depletion under control. This appears to be an area where success is being achieved.[125]

Acid rain

Acid rain occurs when air pollutants, sulphur dioxide and nitrogen dioxide, are emitted by factories and energy plants that use coal and oil. The pollutants combine with water vapour to produce acidic precipitation, or acid rain. Acid rain can and does cross international boundaries, as between the US and Canada, or between European countries. In Taiwan's case, its problem with acid rain is produced by a combination of local industrial processes as well as long-distance transport of acid rain from mainland China. The problem is most serious along the west coast plain, where pH values range from 4.5 to 5.6. Eastern Taiwan, protected by the high mountain interior from prevailing winds from the west, does not have a serious acid rain problem.[126] Obviously, Taiwan can not completely solve this problem without cooperation from China, which has huge environmental problems that make Taiwan's pale by comparison.

Risks from nuclear power

Nuclear power, born out of World War II, was initially seen as a lifesaver for energy-hungry nations in the years shortly after the war, because it promised relatively cheap energy that would not pollute the air or water, and could help lessen dependence of nations on expensive or undependable supplies of imported coal, oil, or natural gas. Thus, nuclear power plants sprang up in many parts of the world, especially Europe, the US, the wealthier countries of Asia, and so on. The plants were not cheap to build, and were technologically complex, but once built were cheaper to operate than plants run on fossil fuels. After a couple of decades of experience with nuclear energy, however, the world started to better understand the drawbacks to such energy. For one thing, the plants produced large quantities of extremely hazardous nuclear waste, and no one really knew what to do with it, since it takes more than 300 years for low-level radioactive materials to decay, and at least 100,000 years for high-level wastes.[127] Thus, nations discovered that they had to find very safe, essentially permanent storage sites for this deadly waste, and no one wanted such sites near them, even if they could be designed to be foolproof (the proverbial NIMBY, or 'Not In My Back Yard', syndrome). In addition, nuclear plants are only as safe as the people operating them, and people make mistakes, as happened at Three Mile Island in Pennsylvania in July 1985, and at Chernobyl, in Ukraine, in April 1986. Those two events, more than any others, severely undercut public support for nuclear power around the world. The Chernobyl incident was the deadliest anywhere, so far, and provided chilling proof to critics of nuclear power of the very real dangers involved. Even though Chernobyl was not in a highly densely populated area (about 100 km from Kiev), more than 100,000 people had to be evacuated from 91 settlements in the region, many of them permanently. About 8 per cent of the land near Chernobyl was permanently contaminated, and a 30 km zone around the plant was classified as highly radioactive. More than three million people were victims of radioactive poisoning to some degree. Neighbouring Belarus was also affected, including 100,000 children who had to be sent away for protection from radiation.[128] Chernobyl made many people around the world seriously question whether nuclear power is worth the risk, especially in highly populated countries such as those in East Asia.

Taiwan falls in that category. Taiwan also was once illustrative of the same kind of political environment as the former Soviet Union, Eastern Europe, or China, in which critical decisions such as building of nuclear power plants were made by autocratic governments without any public discussion or dissent allowed. Thus, Taiwan got into the business of nuclear power during the KMT era, when the government was solely focused on industrializing the island as rapidly as possible (see Chapter 1). That required large amounts of energy, and the island had almost no domestic energy resources except some hydropower and negligible coal and natural gas deposits. Moreover, thermal power plants produce various air pollutants, including CO_2 and NOx, among others. Nuclear power was seen as a partial solution to the problem, and had the KMT stayed in power the government probably would have proceeded to build a great many nuclear plants around the island, irrespective of the risks. As it was, the KMT managed to get three plants (with a total of six reactors) constructed while they still had the ability to push them through without dissent: the first plant was built in 1961 at Chinshan in Taipei County, the second in 1974 at Wanli also in Taipei County, and the third in 1978 at Hengchun in Pingtung County in southern Taiwan[129] (see Figure 1.9). By about 1980 the island was getting one third of its electricity supply from nuclear power. That share has since declined to about 18 per cent today because of the increase in construction of thermal plants in the years since (nuclear power's share of total energy is even less, at about 7–8 per cent now).

In 1980 the government initiated plans to build a fourth nuclear plant, in Kungliao township in Taipei County, to help provide further power for the voracious energy appetite of the Northern Industrial Region, centred on Taipei, the key economic region for the whole island. Unfortunately for the government, however, KMT political power was beginning to slip away by the 1980s, as some legislators and some in the public began to have the courage to confront the KMT-run government. Thus, through the 1980s and 1990s, the Executive Yuan and the Legislature tussled back and forth over budget and plans, with the plant still not started. The green light was finally given and construction begun in January 1998, almost 20 years after first being proposed. However, with the election of the opposition DPP government, under Chen Shui-bian, in 2000, the project was again targeted for rejection, as part of the DPP's platform of anti-nuclear power.[130] Unfortunately for Chen, the Legislature was still controlled by the KMT, and thus a political firestorm arose when he tried to halt construction, which was about one-third done at that time. Taiwan's highest court then ruled that Chen's government had erred in the decision. Recognizing the need for compromise with the KMT-dominated legislature if the DPP hoped to get anything else accomplished in his administration, Chen agreed to restart the project in 2001, with the proviso that Taiwan would gradually phase out all nuclear power, including closing the three already operating plants, over the next 50 years.[131]

The troubles for the government were not ended, however. The issue of nuclear waste still was not resolved. During the KMT era, about 40 per cent of the waste was stored on site at the three nuclear plants. Some 97,000 barrels of the waste were also shipped to the remote island of Lanyu (Orchid Island), off the south-east coast of Taiwan and home of the Yami aboriginal tribe, who were hardly keen about this decision but powerless to stop it. In 2002, however, the government announced plans to remove the nuclear waste from Lanyu.[132] The problem is where to put it. One partial solution that was actually explored would be to ship the waste to other countries willing to accept it (for a price, of course), such as North Korea (!), Russia, the Marshall Islands, and the Solomon Islands. While the governments of those places were willing to listen to Taiwan's business proposition (money talks for cash-starved countries), the residents and environmentalists in those places were not exactly welcoming of such an idea, although Taiwan has not yet abandoned that approach. The government also thought of building a permanent depository, for up to 160,000 of the total of 200,000 barrels of waste from all three plants, on Wuchiu, one of the Matsu Islands group just off the south-east coast of China, a plan understandably hotly protested by residents of Wuchiu as well as the PRC government.[133] Now, following on the recommendation of the International Atomic Energy Agency (IAEA), Taiwan's own Atomic Energy Council (AEC) is hoping to build a permanent site somewhere on Taiwan Island itself, in the belief that international standards for such sites (through the IAEA) will guarantee the site's complete safety. Various sites on Taiwan are currently being investigated, and the prospective locations were to be announced by June 2006 (this goal was not met, however), and the final site chosen by November 2008. Construction of the storage facility will take an estimated five years after that.[134] Hence, that means it will be at least 2013 before Taiwan will approach closure on the problem of nuclear waste disposal. Much can happen before the construction even begins, including a change of government in Taiwan, and possibly a government that swings in the other direction, more in line with other Asian countries, such as Japan, South Korea, and China, that are increasingly coming back to favourable views of nuclear power. Dangerous or not, nuclear power may well be the wave of the future for energy production in Asia. Will Taiwan join the bandwagon again (see Chapter 3)?

Conclusion

Taiwan has come a long way in economic development in the last half century, and simultaneously run the gamut of environmental problems as a result. The island's experience thus parallels that of many economically successful nations around the world. Yet, Taiwan has deviated somewhat from the industrial democracies at least, in that there was a time lag, a period of several decades, when critical decisions were made affecting the environment, in industrialization, sewage and waste disposal, energy production, land use, treatment of fauna and flora, and others, that caused greater damage to the environment than probably otherwise would have occurred had there been effective checks and balances right from the start. Fortunately, however, those checks and balances finally started to be put in place by the late 1980s, as Taiwan democratized. Thus, Taiwan is now on a parallel track with other major industrial states, such as Japan and the United States, in seeking solutions to the environmental problems created over the past half century. Progress is already measurable in some areas; in other aspects of the environment Taiwan has yet to really begin. The fundamental issue is quality of life, and how to best preserve and strengthen that quality. The following chapter further explores the efforts under way.

3 Back from the brink

Resolving Taiwan's environmental problems

The experience of most countries since the Industrial Revolution started in the nineteenth century has been for states to strive as hard as they could for industrial (manufacturing) growth, regardless of the cost, because it is industrialization that brings the fastest and greatest economic benefits – in job creation, wealth generation, and national power. The cost, however, is often high, especially in the early period of the process – in increased socioeconomic inequality, harsh labour and living conditions, and, as observed in Chapter 2, mounting environmental problems, especially in overcrowded, burgeoning cities. An intriguing question is: what finally causes a state to recognize and begin to tackle the costs of success, especially in the environmental realm? As one looks at the history of the 'successful' nations, i.e., those that are now highly industrialized and wealthy, virtually without exception one can see a combination of public activism (environmentalism) and government programmes being responsible for the turnaround. Who takes the lead, however? In most cases, it is public activism that finally prods the government into establishing agencies and enacting laws for environmental clean-up and protection. That has certainly been the experience of the United States and Japan, and Taiwan has followed in their footsteps, including having close liaison with the national environmental protection organizations in those two countries, especially the US. It is significant, but not really surprising, that public activism tends to take the lead in those states with democratic systems where environmental abuses can be subject to public scrutiny and criticism. Conversely, one has only to note the experience of the former Soviet Union and the former satellite states of Eastern Europe to see the dreadful environmental consequences of unchecked governmental control of the development process. Russia and the former members of the defunct Soviet Empire are now struggling to come to grips with their horrendous environmental problems. China likewise is now feeling ever more the environmental impact of its inherently unsustainable and contradictory system of political authoritarianism but increasingly free market economy. Unfortunately, China does not yet have a system that makes public activism a very useful tool for moving the government in the direction of environmental protection.[1]

In Taiwan's case, there is general agreement among observers that the environmental movement really began in the 1980s, and that the watershed point came in 1986–7, just as martial law was being lifted and democratization got under way, and the government finally began to get really serious about halting environmental abuse and correcting the wrongs already done (also see Chapter 1). Taiwan thus has had only about two decades since then in which to fix things, so the record is still far from complete and the environmental improvements still imperfect. But Taiwan definitely is moving forward, and there is reason for guarded optimism on the environmental front.

Environmentalism and the environmental movement

The environmental 'movement' implies a concerted, organized effort to correct environmental abuses, and as such did not really exist until the mid-1980s in Taiwan. However, this movement did not suddenly appear out of nowhere. The foundations for it had been laid over a number of preceding years, as individuals and small groups of citizens, especially non-governmental organizations (NGOs), raised their voices against this or that policy or environmental problem. Academics, scientists, and writers played an especially important role in that early period of activism.[2] Members of the political opposition also played a role of course, especially as the DPP emerged as the dominant opposition political party in the 1980s, with its formal founding in 1986.[3] These people and groups helped to spread a growing sense of environmental consciousness among a public generally either too complacent or too fearful to openly express their opinion in public forums. Scholars who have studied that era discern the growing activism and environmental consciousness at that time to have focused around three primary concerns: nature conservation, pollution, and nuclear power.[4]

Nature conservation

In the realm of nature conservation, the first glimmers of an environmental consciousness began to appear in the 1960s through the work of a few natural scientists and others concerned about the flora and fauna of Taiwan and the assault being directed against them (also see Chapter 2). At the same time, NGOs were also key actors in the nature conservation movement, especially groups such as the Wild Bird Society, Homemakers' Union and Foundation, Green Consumers' Foundation, the Society of Wildlife and Nature, and the Nature Conservation Union. This awareness accelerated in the 1970s and eventually jelled around four major conservation debates in the first half of the 1980s: the fate of one of Taiwan's last remaining mangrove forests, in the estuary of the Tanshui River near Taipei; the illegal excavation of ponds for abalone culture along the northeast coast of Taiwan; the plan to construct a second cross-island highway in central Taiwan that would severely impact one of the largest remaining wild areas of the island surrounding Yushan, the highest mountain in Taiwan (and East Asia); and the plan to construct a dam and a cement plant in Taroko National Park in Eastern Taiwan, one of the island's greatest natural treasures and a prime attraction for tourists visiting Taiwan. There was considerable coverage of these four issues in the local press, which played a key role in stimulating public debate that eventually forced the government to change its plans or to take action to stop potential environmental abuse: the mangrove forest was saved, further building of abalone ponds was stopped, the cross-island highway was re-routed, and the plan for a dam and cement plant was abandoned. Simultaneously, in the 1980s the government, under pressure from the public, emerging NGOs, and academics and scientists, began to establish a system of national parks and nature reserves that have grown in numbers over the years and now set aside large chunks of Taiwan's precious natural environment for future generations (see page 121).

Pollution

In the realm of pollution, activism was very much a localized affair in the early period of environmentalism. Local residents and their organizations targeted specific or suspected causes of pollution. Academics, college students, and writers then often got involved and helped to foment broader public support for this or that aggrieved community, laying the foundation for the anti-pollution movement as a whole that has since swept the island and is now a potent force. A total of over 1,200 local anti-pollution protests reportedly took place between 1980 and 1996, and most of these after martial law was lifted in 1987.[5] Initially, protests tended to be reactive, but in the post-martial law era the protests became both preventive and pre-emptive. Three-quarters of protests took place in northern and southern Taiwan, focused around the two key urban/industrial centres of Taipei and Kaohsiung, respectively, where most polluting sources were located. The primary focus of protests was industrial plants and corporations, but local government also was targeted. Only a small portion of cases led to the creation of civil society organizations (NGOs) that outlasted the immediate protest objective. In other words, once the immediate local problem was resolved, or at least was being addressed by authorities, the local protest effort tended to dissipate.

Among the most well-known of these protests, the following are illustrative:[6]

1 The 'Lukang Rebellion' of 1986–7: This incident is regarded by many as the real start of the environmental movement, and became so famous it was dubbed a 'rebellion'.[7] This case achieved such notoriety, in fact, it became the focus of a significant book in the environmental literature.[8] Residents of Lukang City (in Changhua County along the central west coast), an historic town that played a very important role in Taiwan's early settlement and development, successfully blocked construction of a titanium dioxide plant by DuPont that would have allegedly caused great harm to the water and local fisheries. The DuPont plant would have been one of the largest foreign investments in Taiwan up to that time, and thus opposition to the plant represented a major challenge to the KMT government. Fortunately,

DuPont officials were wise enough to discern the risks involved and withdrew the project voluntarily.

2 The San Yu Chemical Plant closing of 1986: In April, 1986, Taichung County citizens formed an anti-pollution committee to address the problem of the San Yu Chemical Plant, which was emitting hazardous emissions. The public protests eventually got the plant closed down.

3 The Taiwan Cement Plant, 1986: Taiwan Cement, a private enterprise in Kaohsiung, was a major contributor to that city's foul air. Public protests, including a hunger strike by Chu Hsing-yu, a DPP Kaohsiung City councillor, got the company to burn a cleaner fuel grade and reduce hazardous emissions into the air.

4 The Fifth Naptha Cracker: The government-run Chinese Petroleum Company announced in 1987 that it planned to build Taiwan's fifth naptha cracker as an expansion of its petrochemical complex in Houchin, in Kaohsiung City. The residents of the area were not happy at this news and organized a 'self-salvation' association to fight its construction. Protests were held locally as well as in Taipei. The KMT government effectively used persuasion and promises to divide the opposition into two camps – complete opponents and those willing to compromise with the government. The end result was for the protest movement to lose steam, and the plant was finally started in 1990. Thus, in this case, the government still got basically what it wanted but with modifications, so the residents also gained.

5 The Sixth Naptha Cracker: The private petrochemical giant, Formosa Plastics Group (FPG) proposed in the mid-1980s to build Taiwan's sixth naptha cracker in the city of Ilan, in NE Taiwan. The plant was strongly endorsed by the central government, and FPG actively courted local county government and residents to win them over, arguing the importance of the plant to the local area (especially employment), the petrochemical industry, and hence Taiwan's future. Opposition, however, remained strong enough to finally discourage FPG and cause it to withdraw the plan. FPG then took the plan and proposed it to Yunlin County in 1991, where it was warmly welcomed. Why the different outcomes? In Yunlin, industrial job opportunities were much less at that time, and the DPP was not politically strong there yet, unlike in Ilan. The KMT political machinery was able to overcome local opposition.[9]

Nuclear power

As already briefly introduced in Chapter 2, nuclear power has been highly controversial in Taiwan since the island's first reactor went into operation in Chinshan township on the coast north of Taipei City in 1961. By 1985, six reactors at three nuclear power plants accounted for 33 per cent of the island's electricity generation, the peak for nuclear power's share of electricity output (it is now less than 18 per cent). Plans were being laid for a fourth plant, on the north-east coast, in Ilan County. Taiwan was heading for a nuclear future, in the face of a near-absence of alternative domestic energy sources other than some hydropower. The state-owned monopoly, Taiwan Power Company, was moving ahead virtually without any public discussion, until the 1980s changed all that. A public forum in mid-1985 organized by the Consumers' Foundation, the first social movement NGO (founded 1980), forced Taipower officials to open the process to public scrutiny and debate, compelled in part by the controversy that arose after a fire broke out at the third nuclear plant near the southern tip of Taiwan.[10] This was not the only nuclear incident in Taiwan in that period. In the early 1990s it came to light that during the early 1980s steel rods contaminated with cobalt-60 had been widely used in construction beams for various buildings throughout Taiwan, exposing thousands of civilians to excessive radiation. Intense controversy and public outcry followed, including lawsuits.[11] Hence, the anti-nuclear movement also dates from the mid-1980s.

From 1986 onward, the anti-nuclear forces grew in strength and activism, led by middle-class liberal intellectuals, such as Edgar Lin and Chang Kuo-lung (both of whom went on later to head the TEPA) and concerned environmental groups, especially the Taiwan Environmental Protection Union (TEPU), established mostly by university professors and students (see below). Since there was little that could be done about the existing three plants, the anti-nuclear movement joined forces against the fourth nuclear plant, proposed for Kungliao township in Taipei County (see Chapter 2). Because the KMT was slowly losing power and the ability to suppress dissent through the 1980s, the Executive Yuan and the Legislature, increasingly containing members of the DPP, whose platform was in part devoted to elimination of nuclear power on Taiwan, duelled over the proposed plant, forcing repeated postponements. Referendums were held and the results showed increasing opposition to the nuclear plant. Nonetheless, because the government was still dominated by the KMT, the plant was finally given a green light in January 1998, almost 20 years after first being proposed. Then, when Chen Shui-bian was elected as President in 2000, he tried to halt construction of the plant, but the KMT-controlled legislature balked, in part because the plant was already one-third constructed. Taiwan's highest court ruled in favour of the plant, and Chen's government backed down, recognizing the need to compromise with the Legislature. More radical members of the DPP and other anti-nuclear activists accused Chen of selling out and abandoning DPP principles, but Chen countered that he was only being politically realistic. Regardless, the anti-nuclear movement has not given up, and activists continue trying to stop construction of the fourth plant, which at that time was scheduled for completion in 2007, pointing out that the plant site on the north-east coast is highly vulnerable to earthquakes and the impact of future tsunamis.[12]

NPOs and their role

The non-profit sector has been a growing presence in the environmental movement in Taiwan. There actually are several ways to classify the non-profit sector. Collectively, all the organizations in this sector are known as non-profit organizations (NPOs). One study of the non-profit sector in the 1990s differentiated it in two ways.[13] One classification could be on the basis of the non-social movement sector (NSMS) and the social movement sector (SMS). The majority of NPOs started out in the NSMS, while the SMS is the newer part and played an especially important role in the democracy movement of the 1980s and onward. In the 1990s, however, the two parts, NSMS and SMS, tended to converge to some extent in efforts to reform Taiwan society. The other possible classification is between membership associations (commonly referred to as non-governmental organizations, or NGOs) and endowment-centred foundations. By the late 1990s, there were nearly 11,800 NGOs formally registered (excluding trade and political organizations), the vast majority founded in the 1980s and 1990s. These many NGOs were devoted to a wide range of causes, including the environment. At the same time, there were almost 1,600 foundations of all types, the vast majority also in existence only since the 1980s, 70 per cent private and 25 per cent corporate, the rest mixed, and just as diverse in purpose as the NGOs. Of those foundations, 32 were devoted to some aspect of environmental protection; 3 of the 300 largest were involved with the environment. In actual functioning, however, the NGOs and foundations are really different only in legal definition. Another study of the non-profit sector, also done in the late 1990s, showed very similar results.[14] Today, in the literature one commonly sees the terms NPO and NGO used interchangeably, even though some of these organizations are government-sponsored.

From the perspective of the environment, the TEPA maintains that there currently are a total of 104 NPOs related to the environment, ranging from small grassroots organizations to bigger government-sponsored research and development institutes. Of these NPOs, 29 were registered with the TEPA and 75 with the Ministry of Interior. These 104 NPOs can be divided into four groups:[15]

Government-sponsored organizations

Since the late 1980s, the government has established a number of non-profit research and development organizations devoted to environmental issues. These NPOs work closely with the government or corporations, offering advice on policies and promoting cleaner production practices. Prominent among these NPOs are the National Centre for Cleaner Production and the Institute of Environment and Resources, which collaborate with the government and civil society organizations in developing environmental protection strategies. Also prominent is the Industrial Technology Research Institute, which administers the Centre for Environmental Safety and Health Technology.

Civil society organizations

Civil society organizations that are very active in the field of the environment can be divided into two types: natural resource preservation and nature conservation. Prominent among the former are the Homemaker's Union and Foundation, and the Beautiful Taiwan Foundation, both of which try to educate the public and government on issues such as recycling, reduction of domestic waste, water and energy preservation, and others. Examples of the latter type include the Society of Wilderness and the Taiwan Ecological Research Network, both concerned with the preservation of Taiwan's natural habitats and ecosystems. Some NPOs are advocacy-focused, such as the Green Citizen's Action Alliance, a grassroots movement dealing with anti-nuclear and anti-waste issues.

Animal protection organizations

Many NPOs specialize in animal protection. Some focus mostly on helping domestic animals, such as the Help-Save-A-Pet Foundation Taiwan, or the Taiwan Association for the Protection of Animals, which work to rescue stray animals and provide shelter and care for them. Many other NPOs focus on the protection and study of wildlife, including the Association for the Protection and Breeding of Rare Animals, or the Wild Bird Society of Taiwan. These NPOs bring together researchers and public, and now are found in almost every corner of Taiwan. An estimated 4,000 or more of these NPOs deal just with entomology. A few of the NPOs deal with international issues, such as the Society for Wildlife and Nature (SWAN), founded in 1982, that has evolved into an international organization supporting many projects worldwide but especially in Asia and Africa. For example, SWAN played an instrumental role in a number of important initiatives, starting in the early 1980s, to protect Taiwan's precious wildlife. Likewise, Taiwan's unique birds, including the Mikado Pheasant, would be in much worse shape today if not for the efforts of the Wild Bird Society.[16]

It should be noted that the precise number of NPOs, whether devoted to the environment or not, fluctuates over time, as smaller, weaker organizations come and go. The proliferation of these NPOs over the past 15–20 years, nonetheless, is a powerful demonstration of the flowering of Taiwan's civil society and political democratization.

The Taiwan Environmental Protection Union

The Taiwan Environmental Protection Union (TEPU) is one of the most important NPOs in Taiwan and illustrative of the role of the non-profit sector in protecting the environment.[17] Founded in November, 1987, the TEPU has about 1,000 members drawn from university professors, schoolteachers, medical doctors, homemakers, students, workers, farmers, etc., in other words a broad cross-section of contemporary society. The headquarters are in Taipei, and there are ten local chapters around the island, in the counties of Taipei, Ilan, Hualien, Taoyuan, Changhua, Yunlin, Tainan, Kaohsiung, and Penghu. Seven counties do not yet have a branch, but members of the TEPU are found in every locale on the island. In the span of almost 20 years' existence, the TEPU has become powerful and influential, and often involved in controversy. In many ways, the

TEPU illustrates the power of NPOs in Taiwan in their potential ability to influence social change and policies. The founders of this NPO were highly educated and informed citizens who had chafed for many years at the straightjacket imposed on Taiwan society by the KMT-controlled government and the environmental abuses that had resulted.[18] They deplored the degradation of the island's natural environment. The ending of martial law in 1987 gave this group the green light to charge ahead and formally organize, so that they could begin to openly and aggressively help change the situation in Taiwan. Before the 2000 election, however, the KMT-controlled government refused to let the TEPU formally register with the Ministry of Interior because of the organization's name (which the KMT wanted to be ROCEPU).

Much of the TEPU's activities, especially in the early years, were focused around public protests, but in more recent years the TEPU has also engaged in a variety of positive programmes of environmental education. In other words, the TEPU wanted to be seen not just as a negative force opposing this or that, but also as a positive force providing alternative approaches to development and the environment. In some respects, the TEPU has become like a non-governmental shadow organization promoting many of the same things as the government-run Taiwan Environmental Protection Administration (TEPA), the main central government body overseeing the environment (see page 95). The two organizations have some remarkable similarities in their goals and activities, if not in their funding and organizational structure. For one thing, unlike the relatively well-funded TEPA, the TEPU relies primarily upon donations and grants to fund its diverse activities, whereas the TEPA has a government-funded budget in the hundreds of millions of dollars.

The three over-arching goals of the TEPU are (1) to create a nuclear-free environment; (2) to diminish and prevent pollution of Taiwan's land, water, and air; and (3) to protect biological diversity, i.e. nature conservation. In other words, the TEPU has set an ambitious and broad agenda that covers just about everything in the environmental sphere. The NPO works toward those three objectives through a variety of activities and campaigns:

1 Promoting environmental awareness: The TEPU publishes a monthly magazine, *Taiwan Environment* (in Chinese); organizes various activities in public education, such as student summer camps, lecture series, and many others; holds conferences, including international ones, about various environmental issues; and participates in international environmental meetings.
2 Pushing environmental legislation: The TEPU lobbies for specific legislation, such as a fundamental law for environmental protection and standards, requiring environmental impact assessments for all projects and policies, protection of soil and water resources and prevention of erosion.
3 Protecting forest, water, and land resources: The TEPU organized the 'Forests Take to the Streets' march to demand the end of cutting of Taiwan's primary growth forests; opposes the spread of construction and

other activities on the fragile slopelands of the island; opposes the expansion of limestone mining and cement production on the east coast; opposes the excessive pumping of groundwater in coastal waters for commercial fish ponds (the main cause of land subsidence along the west coast); and opposes further construction of large dams on the island's rivers.

4 Promoting a nuclear-free island: The TEPU carefully monitors the existing three nuclear plants and disseminates information about them, including holding an annual anti-nuclear march; facilitated the holding of popular referendums on the fourth nuclear plant in Taipei County (see page 71); endeavours to get voted out of office those National Legislators who have supported the various nuclear plants, including the fourth one; joined forces with anti-nuclear activists in other countries to oppose nuclear testing by other countries, and to promote nuclear-free environments everywhere.

In 2000, the TEPU conducted an island-wide environmental survey, involving telephone interviews of more than 1,100 residents of Taiwan. Some of the very interesting results were:

1 Twenty-five per cent thought that Taiwan's environment was in really bad shape, while 20 per cent thought it was just bad, and 50 per cent regarded the situation as 'normal'.
2 The main causes of the environmental situation were seen as: inadequate enforcement or observance of environmental laws (57 per cent), followed by the pro-development attitude of policy makers (33 per cent), then the belief that government-owned and party-owned businesses violate the laws first (15 per cent).
3 A whopping 65 per cent thought the best solution to garbage was to reduce, recycle, and reuse, whereas only 32 per cent favoured incinerators.
4 The growing problem of mud slides and soil erosion on slopelands was seen by 86 per cent as the result of excessive tree cutting and construction on hillsides.
5 In regard to the growing shortage of water in southern Taiwan, the interviewees felt that the best solution was water conservation by all water users (industry, agriculture, households), not construction of new dams in an earthquake-prone region.
6 The growing crush of motor vehicles and their impact were seen as a real problem that could best be solved (74 per cent) by greater emphasis on public transport, not more and wider roads.
7 As for nuclear energy, 42 per cent thought it should be gradually phased out, while 32 per cent viewed it as safe and necessary, and 21 per cent saw it as unsafe but necessary, indicating still somewhat ambivalent feelings about nuclear power and Taiwan's critical energy needs.
8 A solid majority (68 per cent) felt that in terms of Taiwan's economic future, the island should encourage more low-pollution, low energy-consumption, high value-added industries. The current heavy emphasis on

energy-intensive industries such as petrochemicals, steel, pulp, and cement, which use one-third of overall energy but contribute barely 7 per cent of GDP, received very little support.

9 An overwhelming 91 per cent agreed that Taiwan should abide with international environmental treaties and the Taiwan government should do all it can to protect the environment.

In short, one could say that this survey suggests the people of Taiwan have come a long way in developing a sense of environmental consciousness, and they are very much in agreement with most of the policies and viewpoints of the TEPU. For those concerned about Taiwan's environmental future, this is an encouraging trend.

Political parties and their roles[19]

Political parties have had both negative and positive effects on environmentalism in Taiwan over the decades. As already noted in Chapter 1, Taiwan had only one legal, functioning party until the mid-1980s, i.e. the Kuomintang (KMT). Since 1987 a number of other political parties have emerged, some much more important than others.[20] As of June, 2005, a total of 110 political parties had registered with the Ministry of Interior. Most, however, are insignificant in electoral politics. As of 2004, the top five in membership numbers were: the Democratic Progressive Party, Kuomintang, People First Party, Taiwan Solidarity Union, and New Party. Following is a brief description of each of the top parties and their environmental positions. In examining the parties, it is necessary to distinguish between party platforms and the actual position of party members toward the environment. Unfortunately, the recent trend has been for all of the parties to become basically the same in their outlook toward the economy, at least in terms of putting economic growth above the environment. This undoubtedly is partly a consequence of the difficult economic times that have faced Taiwan since the Asian fiscal crisis of 1997–8 and the sluggishness of economic growth around the world since then. This is not unusual among states around the world, of course. Growth and jobs almost always tend to have higher priority than the environment, everywhere.

Pan-Blue (Fan Lan) Alliance

Kuomintang (KMT or Nationalist) Party

The Kuomintang[21] (which means literally 'National People's Party', usually referred to as simply as the 'Nationalist Party') is the oldest and once-dominant party on Taiwan. It celebrated its 100th anniversary in 1994, having been founded in the waning years of the Qing Dynasty on the mainland by revolutionaries, including Sun Yat-sen. Brought to Taiwan in 1945 with the defeat of Japan, the KMT completely dominated the government and political system until the 1980s, in an authoritarian one-party state that has been likened to a Leninist system (see Chapter 1). The KMT today is a very different entity, however, having undergone many reforms and changes to meet the challenge of opposition parties and a true democratic system. The changes began under Chiang Ching-kuo, who was President of the ROC from 1978 until his death in 1988. He was succeeded by Lee Teng-hui (then Vice President), a native Taiwanese (Hakka by ethnicity) but a member of the KMT, who served as President until the year 2000, when the current President, Chen Shui-bian (of the DPP) took over. Hence, the government was still effectively controlled by the KMT through the decade of the 1990s, albeit with diminishing power. Although a KMT member, Lee was important in serving as a bridge between the native Taiwanese and the Mainlanders and in promoting political democratization. Lee actually supported many of the policies of the DPP and other opposition groups, including being a strong advocate for the environment.[22] Hence, the 1990s was a transition period in many ways.

As for the environmental position of the party, the KMT had little to say about it during the decades of absolutist rule, when industrialization and economic development at just about any cost were the overarching objectives.[23] Nonetheless, the KMT did not totally ignore the environment. For example, the government launched a pilot study for environmental impact assessments in 1981,[24] and the Environmental Protection Bureau (precursor to the current TEPA) was founded in 1982, as was the National Park Service. More recently, however, with the formation of opposition parties and having to defend its position in public campaigns for offices, the KMT and its members have had to stake out a more explicit environmental policy in order to court voters. That position could be summed up as being pro-environment but only insofar as protecting the environment does not do damage to the economy and thereby in any way weaken Taiwan, economically or militarily, given the grave threat from the PRC. That position also used to include a pro-nuclear energy policy, with the argument that Taiwan could not afford to be anti-nuclear in light of its need to otherwise import large amounts of coal and oil for energy production. Only toward the end of the 1990s did the KMT finally come around to supporting the eventual phasing out of nuclear power on Taiwan in coming decades, but that stance could well flip back to support of nuclear power in the future (see page 71).

The KMT's position on the environment today is summed up in its 'New Proposals for Development of Taiwan', which contain a section titled 'To initiate economic reconstruction for sustained (sic) development to provide a better life for all', of which just two statements focus on the environment – 'Resuming the comprehensive national land development programme to better manage natural resources' and 'Promoting urban renewal, ecological preservation and tourism'. In other words, one can see some effort to move the party in the direction of environmentalism, but the main thrust of the platform is still strongly pro-business and pro-growth.[25]

People First Party (Ching Ming Tang)

The People First Party (PFP) was born out of the 2000 presidential election, in which former KMT provincial governor James Soong ran as an independent and was just narrowly defeated by the DPP's Chen Shui-bian. Soong was formerly a KMT member but then had a falling out with the party and thus ran as an independent. In the 2000 election, Lien Chan (KMT) received 23 per cent of the vote, Soong 36.8 per cent, and Chen (DPP) 39.3 per cent. Had Soong not run as an independent, the KMT would have retained control. After the election, his supporters urged him to form his own party in order to build a stronger base; hence the PFP was born, which has since become the third largest party after the DPP and KMT in the Legislative Yuan. In the 2004 Presidential Election, as well as the November 2004 Legislative Election, the KMT and PFP, sharing similar platforms and views, especially about the desirability to defeat the DPP, formed an alliance known as the Pan-Blue.[26] While not a formal merger, the alliance almost worked in the 2004 Presidential Election, in which the Pan-Blue alliance garnered 49.89 per cent, while the DPP and Pan Green alliance just barely topped them with 50.11 per cent, a razor-thin margin. The Pan-Blue camp was even more heartened by the legislative election that took place in late 2004, in which the Pan-Blue camp managed to win 114 seats of the 225-member Legislature, while the Pan-Green alliance (see below) got only 101 seats. Moreover, the KMT gained 11 seats from the 68 it won in the 2001 elections, for a new total of 79, while the PFP suffered a loss of 12 seats, to just 34, down from 46 in 2001. These results gave the Pan-Blue camp 49.81 per cent vote share, compared to 46.26 per cent for the Pan-Green camp.[27] In short, Chen Shui-bian's hopes for gaining control of the Legislature and implementing his reform agenda were dashed. Observers seemed to agree that the strong showing of the Pan-Blue camp was evidence of the public's nervousness about Chen's intentions regarding independence, with a majority of the public preferring to keep the status quo for the time being and doing nothing to provoke the PRC,[28] as well as due to over-confidence by the DPP.[29] These election results were not only bad news for those supporting Taiwan independence, the outcome also raised concerns of environmentalists.

New Party (Hsin Tang)

The New Party was founded in 1993 during the Lee Teng-hui Presidency by young Deep Blue (Mainlander) people who disliked Lee. The founding members were strongly pro-environment, except for nuclear power, and included Chao Shao-kan, who had just served as TEPA head (see page 96) and became the first party chair, and Hau Lung-bin, who would later also become head of the TEPA during Chen Shui-bian's first term. The New Party never managed to really capture the voting public's support, however, and sputtered along as a relatively minor party, and even lost influence in the 2000 and 2004 elections. In the latter elections, in fact, the New Party joined forces with the

KMT under Lien Chan in the Pan Blue alliance, but won only a few seats. The party's influence today is not great.

Pan-Green (Fan Lu) Alliance

Democratic Progressive Party (Min Ching Tang)

The Democratic Progressive Party (DPP), originating out of and drawing its strength from the native Taiwanese, was formally born in September, 1986, but had been gestating for some time before that. Indeed, the DPP grew out of the loosely organized opposition in the martial law era known as the *Tang-wai* (lit. 'outside the party', i.e. outside the KMT). Originally created by family members and defence lawyers of political prisoners who were willing to defy martial law and confront the KMT-controlled government, the DPP gathered strength year by year and grew into the main opposition party. Over the nearly 20 years of its existence, the DPP was in the forefront pushing for political reforms, including the end to martial law in 1987, the general elections in 1992, and the first direct presidential election in 1996.[30]

The DPP from the very beginning was strongly pro-independence, i.e. arguing for a Taiwan free of both PRC (Communist) and Kuomintang rule, which meant in effect a 'Republic of Taiwan' or 'Republic of Formosa' with its own distinctive flag that is still the banner of the DPP.[31] Over the years, however, especially since Chen has been President, the DPP leadership has been forced to recognize the dangers of its pro-independence stance, and taken a more muted position that accepts the necessity to reach some kind of accommodation with the PRC and to have the KMT as probably a permanent competing opposition party. Chen has put formal independence on the back burner, while carefully maintaining the 'Republic of China' name for the government.

A key position of the DPP, whose party colour is green, is attention to the environment. Indeed, as the environmental movement arose in the 1980s, it became closely linked with the DPP, which saw environmentalism as not only a worthy cause for the welfare of Taiwan's people but also likely to be helpful to the DPP in winning votes. This alliance has been part of the reason for the gradual withering of support for the KMT, with its continued strong backing of growth-oriented industrialization. Thus, environmentalism and environmental groups were an active creator of the democratic transformation in Taiwan, and an essential part of the DPP's success. However, some of the environmental activists became increasingly disillusioned over what they regarded as the DPP's too conservative views on environmental issues, as the DPP fought the 1996 campaign and realized the need to move closer to the middle of the road if it wished to win. Thus, these more determined environmentalists bolted the DPP and formed their own party, the Green Party Taiwan (see page 88). Nonetheless, the environment remains a cornerstone of the DPP platform. That platform states:[32]

> Protect the environment: the government should revise its policies and give priority to preserving the ecological environment and upgrading living standards rather than only focusing on industrial development. The government should make careful assessments of policies that pertain to the development and utilization of natural resources, production and utilization of chemicals and construction of nuclear power plants.'

The DPP thus maintains in its public relations that it is for sustainable development, environmentally sound industrial policies, support of international environmental agreements (such as the Kyoto accord on global warming), and a totally non-nuclear Taiwan. The opposition argument that the DPP is anti-business is countered by the DPP that it is not anti-business, but it is against polluting, irresponsible businesses, especially the KMT-founded and KMT-run enterprises and national government monopolies in key industries such as railways, domestic banking, shipbuilding, petroleum refining, tobacco and alcoholic beverages, electric power, sugar, salt, and steel. As proof of its support for industry, the DPP-led government came out with a slogan to promote Taiwan as a 'Green Silicon Island', in reference to the silicon chip that plays such a critical role in the IT industry and economy. The idea is that Taiwan must continue to pursue economic development, especially focused on IT, but must do so within the framework of a sustainable environment.[33]

Some environmentalists remain unmollified, however. They complain that the DPP lacks the will to ensure the nation's sustainability, pointing fingers particularly at what they call the 'Ten Nightmares', or key national development projects or proposals, which they oppose but the DPP-led government is implementing or proposing to implement.[34] Some of these projects predate the DPP government, of course, but the Chen government is on the horns of a dilemma in what to do with these so-called 'nightmares' (see page 117).

Taiwan Solidarity Union (Taiwan Tuan-chieh Lien-meng)

The Taiwan Solidarity Union (TSU) is one of the newest additions to the political scene. Founded in July, 2001, the TSU came into existence precisely because of the deep divisions between the KMT and the DPP. The TSU claims to want to 'stabilize the political situation, promote the economy, consolidate democracy, and strengthen Taiwan'. The 'spiritual leader' of the TSU is former President Lee Teng-hui, who argues that the TSU is the party most strongly promoting the path of Taiwanization and should reach out to all of Taiwan's ethnic groups.[35] For the 2004 elections, the TSU joined forces with the DPP, in what has become known as the Pan-Green alliance, since the TSU's positions were closer to those of the DPP than the KMT. The links with the DPP and Lee Teng-hui have given the TSU the public image, though, of being strongly pro-independence, a perception that matches reality. Environmentally, the TSU claims to be for a 'Green Taiwan', but in reality is like the DPP in putting the economy first, environmentalists argue.

Green Party Taiwan

Generally not considered one of the Big Five (now, more properly speaking, the Big Four) political parties, the Green Party nonetheless is a significant minority party created in 1996 by a disgruntled group of members of the DPP, who felt that the DPP was not giving high enough priority to the environment. This, in spite of the fact that the DPP deliberately chose green as its identifying colour, in line with its promotion of a 'Green Island', referring to an environmentally stable island. Unfortunately, the realities of trying to win control of the government caused the DPP to sometimes dilute its pro-environment position for the sake of winning votes or achieving political compromise. The Green Party is unwilling to compromise when it comes to the environment.

Taiwan's Green Party is actually but one of many Green Parties found in countries all over the world. These parties date back to various times in the post-World War II era, especially in Europe, but are also found in North America, Australia/New Zealand, and even parts of Africa. Although each country's Green Party has its distinctive local features, these parties internationally tend to be quite similar in their promoting four basic principles: ecological sustainability, grassroots democracy, social justice, and world peace. Although relatively small in numbers of adherents, Green Party members tend to be strong activists who make up for their smaller presence by vigorous demonstrations, protests, and programmes.

The Declaration adopted by Taiwan's Green Party in its inaugural General Assembly in June, 1996, makes clear the long-range objectives:[36]

> The Green Movement entails ecological, environmental, peace, feminist, and human rights movements. The Green Movement is the manifestation of citizens who have risen to oppose the environmental pollution and social crisis brought about by industrialism. Technological developments, human shortsightedness, and the blind pursuit of economic growth since the Industrial Revolution lead to forest destruction, water pollution, acid rain, global warming, and a myriad of ecological crises. International arms races and pollutant exports further exacerbate this crisis and constitute a serious intrusion of the living rights of millions of women, children, and minority groups. We human beings are confronted with a dilemma that the only way to safeguard social justice and maintain survival conditions is through mutual cooperation and understanding. The Green Movement is the new social movement in the end of the 20th century.

Taiwan's Green Party started out with a respectable showing in the 1996 national assembly election, in which the party's 13 nominees won an average of 3 per cent of the vote, but the party has yet to win any seats in the assembly. The Green Party is not nearly as influential as the TEPU (see page 80). As is true in virtually all other countries, the Green Party in Taiwan is viewed by a majority of the electorate as an extremist group whose objectives, while perhaps noble in principle, are too utopian to be realistically implemented in contemporary society. Yet, the Green Party appears to slowly gain adherents as more and more people come to realize the non-sustainability of Taiwan's present path of development.

A divided island

Taiwan today is thus a deeply divided island politically. The DPP and the KMT, as the two principal political parties, with their alliance partners, the TSU and PFP, respectively, have fundamentally divergent views on what is best for Taiwan's economy, its environment, and its future relationship with the PRC. This makes for a volatile political climate and highly fragmented, divided legislature and electorate.[37]

Thus, one can see the divisiveness both geographically and ethnically. The Pan-Blue coalition tends to be strongest with the Mainlanders, concentrated in the Taipei and northern Taiwan region, as well as in eastern Taiwan in Hualien and Taitung where there are large numbers of retired servicemen living. The Pan-Green alliance tends to be dominated by the ethnic Taiwanese (especially the Hoklo) and is generally strongest in southern Taiwan, including the Tainan–Kaohsiung area, as well as in Ilan. The Hakka, up to now, have tended to be pro-KMT, interestingly, but now are about 20–30 per cent for the Pan-Green. The aborigines generally are pro-KMT, but this is slowly changing.[38] Taiwanese (both Hoklo and Hakka) who do favour the Pan-Blue tend to come primarily from the business community, perceiving the Pan-Blue as more supportive of business and linkages with China. Some Taiwanese farmers and rural people, especially local elites, also lean toward the Pan-Blue because their power came through the government-run farmers' associations and similar rural organizations in past years. One could crudely generalize that the Pan-Blue's strength comes primarily from those with power and money, the Pan-Green's strength from those with less.

In terms of the environment, this highly charged and divisive political climate makes it more difficult for the government and various agencies charged with protecting and cleaning up the environment to carry out their duties to maximum effectiveness. To some extent, this situation prevails in almost every country, with pro-environmental forces confronting pro-business, pro-growth forces who argue that priorities have to be maintained, and the highest priority has to be economic growth and jobs. In Taiwan's case, national security, i.e. maintaining the island's defences against China, plays an additional key role that makes the political climate all the more complicated, because a strong defence depends directly on a strong economy. To suggest curtailing the rate or nature of economic growth for the sake of the environment tends to imply a lack of true patriotism, in the minds of some. These are not the optimum conditions in which to carry out environmental protection. Environmental issues were all but ignored, for example, in the 2004 legislative campaigning.

Government and its role

Protection and clean-up of the environment ultimately has to become the responsibility of government, if the task is to be accomplished. NGOs and individual citizens can promote environmentalism, and on their own accomplish much. But there is no substitute for solid public policies enacted and effectively enforced by government agencies at the city, county, and national levels. Political parties run governments, however, and hence the degree to which good environmental policies are enacted and then properly enforced depends very much on the parties doing the governing. As already discussed, Taiwan's governing system has gradually evolved into a multi-party system in the last 10–20 years, and ostensibly pro-environmental parties (the DPP, TSU, Green Party) have been gaining ascendancy but are not yet totally in control, and political circumstances sometimes force parties to compromise with environmental principles. Hence, the history of government's role in environmental protection in Taiwan is very chequered, with some successes but also some failures. To put it another way, one might say that the government response to the environmental crisis is currently in its adolescence and struggling to come of age.

Government structure for environmental protection[39]

Government response to the environmental problems of Taiwan has evolved over several decades. As concern about the environment slowly emerged in the 1970s and 1980s, agencies at various levels began to be assigned elements of the environment as their responsibility. However, in general, in the pro-growth, single-party political setting of that martial law era, no agency at any level really took the responsibility very seriously. Government commitment, especially at the county and 'national' (island-wide) level, lagged well behind rising public concern.

One could say that the year 1987 was a watershed point in the environmental struggle, because the cabinet-level (Taiwan) Environmental Protection Administration (TEPA), the highest authority in the environmental fight, formally came into existence in that year. Hardly coincidentally, that was also the year when martial law ended and the political flowering of Taiwan really got under way. Hence, one can see the impact of that watershed year in the structure of governmental response to the environmental struggle and the specific policies enacted by the various agencies and units of government that came to be involved in the fight for environmental quality.

The TEPA's establishment followed on the heels of a series of measures taken by the ROC government at various levels, starting in the late 1960s and early 1970s, to begin to address environmental problems and issues. In 1974, a series of measures were implemented, including the first acts to deal with solid waste disposal, water pollution, and air pollution. At about that time, also, a pioneer planning group was set up in order to develop an integrated plan for controlling pollution and preserving quality of life. This was implemented in

1979 as the Environmental Protection Scheme. Then in 1982, the Bureau of Environmental Protection under the Department of Health was set up, as were the Department of Environmental Protection under the Taipei Municipal Government, and a similar Department under the Kaohsiung Municipal Government. In 1983, the Provincial Government also established its own Environmental Protection Bureau. As Taiwan's industrial growth accelerated, and the pace of environmental degradation also increased, the government felt the need for greater coordination among the various units and thus established an ad hoc Environmental Protection Committee directly under the Executive Yuan at the ministerial level in 1985. Following establishment of the TEPA itself in 1987, with the upgrading of the Bureau of Environmental Protection, in early 1988 the Provincial Environmental Protection Bureau was also upgraded to a Department of Environmental Protection. Likewise, in September, 1988, the Environmental Protection Bureaus in the various counties of Taiwan were established, as were those in many cities and towns. Thus, by the end of 1988, the formal structure of environmental agencies, at all levels within Taiwan, was more or less in place and these organizations were ready to move ahead more aggressively with fulfilling their responsibilities.

The current organization and lines of authority among the many government units with environmental protection (including resource conservation) responsibilities are shown in Figure 3.1.[40]

National Council for Sustainable Development

At the top is the Executive Yuan,[41] which has a National Council for Sustainable Development (NCSD), an advisory body established in 1997 with the broad goal of incorporating sustainable development concepts into government administration. Established initially in response to the problem of global climate change, the NCSD is an interagency group, whose chief is the Vice Premier, and the Executive Secretary is the head of the TEPA. The NCSD is advisory to the central government, to discuss major issues, and does so currently through nine working groups: Sustainable Vision; National Land and Transportation; Resources and Industry; Biodiversity; Livelihood and Production; International Environmental Protection; Health Risks; Sustainable Education; and Climate Change and Kyoto Protocol Response. As such, the NCSD currently has 37 members drawn from top administrative positions in the government as well as key academics and other public figures (such as heads of leading NGOs) prominent in the environmental arena.[42] One could liken the NCSD to a grand sort of 'think tank' at the highest level, involving leading thinkers and political leaders, trying to integrate and stimulate the many facets of contemporary Taiwan to produce a truly sustainable system for the island. (See also Chapter 5 for further discussion of this important body.)

The primary workhorse in the environmental realm, however, is the (Taiwan) Environmental Protection Administration (TEPA), the only government agency at the national (island-wide) level that is devoted solely to protection of the

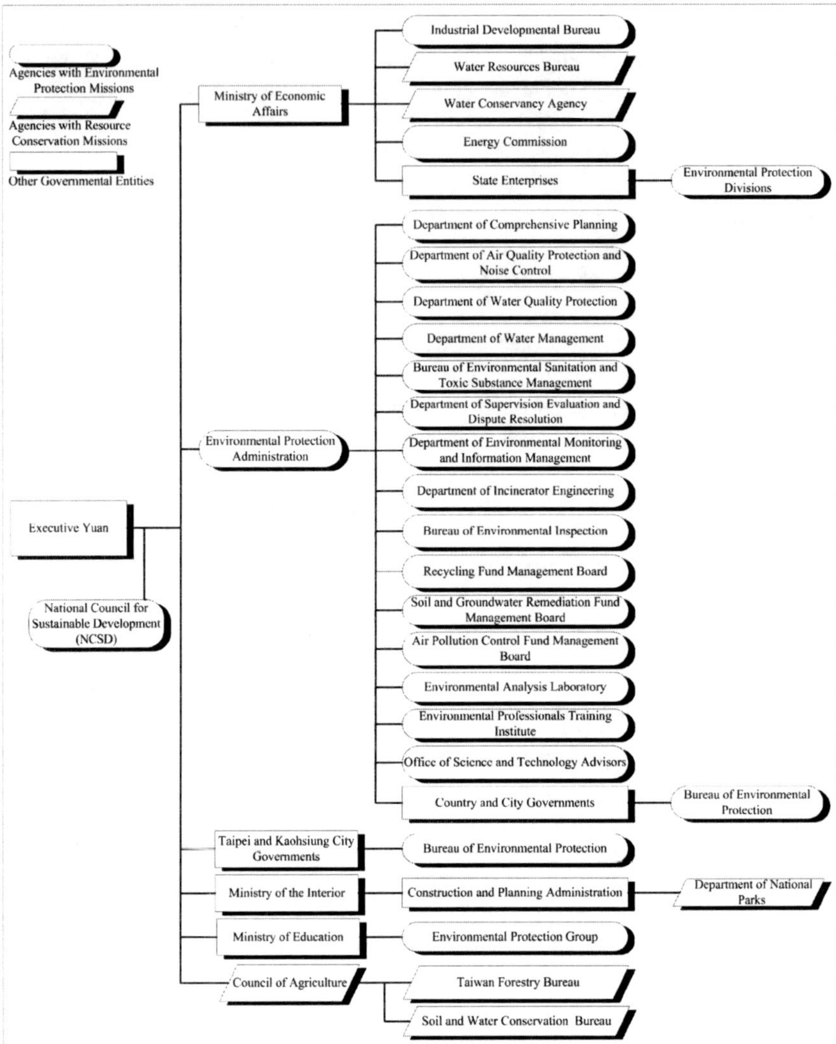

Figure 3.1 Government structure for environmental protection (source: TEPA and other agencies).

environment. As discussed below, the TEPA has a complex mix of branches or sub-units devoted to various aspects of environmental protection. But first, a brief overview of the other agencies and units in the total environmental structure.

Also under the Executive Yuan are a number of other units which, while not devoted solely to the environment, nonetheless do have important sub-units that deal exclusively or partially with environmental matters:

1 The Ministry of Economic Affairs (MOEA) is involved in environmental issues through its Industrial Development Bureau, Water Resources Bureau, Water Conservancy Agency, and Energy Commission. In addition, the state-operated enterprises (SOEs), which are all under this ministry, each have their own environmental protection divisions (for example, the Taiwan Sugar Corporation).
2 The Ministry of the Interior, through its Construction and Planning Administration, administers the Department of National Parks (see page 121).
3 The Ministry of Education has an Environmental Protection Group, whose task is to promote environmental education.
4 The Council of Agriculture (COA) has long had a hand in environmental matters, in its role over the decades of helping promote the well-being of farmers and the agricultural sector.[43] This council also administers the Taiwan Forestry Bureau and the Bureau of Water and Soil Conservation.
5 The Taiwan Provincial Government once had a number of units dealing with the environment, until the Provincial Government began to be phased out in the 1990s and the functions transferred elsewhere.[44] These provincial-level units included: the Department of Agriculture and Forestry, the Bureau of Water & Soil Conservation, the Forestry Bureau, and the Water Conservancy Bureau, all heavily involved in resource conservation activities. The Provincial Government also had its own Department of Environmental Protection. The Council of Agriculture took on administration of the Taiwan Forestry Bureau and the Soil and Water Conservation Bureau, as noted above.
6 Each of the 16 counties in Taiwan, and the five largest provincial-level cities (Chiayi, Hsinchu, Keelung, Taichung, and Tainan), have their own Environmental Protection Bureaus (EPB), responsible obviously for environmental issues within their geographical areas. These offices at the city and county level used to be under the Taiwan Provincial Government, and then after the abolishment of the Provincial Government they were put under the direct control of the TEPA. Altogether, thus, there are 23 EPBs on Taiwan (16 counties + five cities + Taipei and Kaohsiung).

In sum, there would seem to be a well-developed hierarchy of environmental units from the top down, through each level of government administration, to provide the framework for effectively addressing environmental issues. In reality, of course, the structure is only as good as the laws and regulations, and the people running the agencies and units. The historical record reveals that enforcement of environmental laws and regulations, where they exist, has often been spotty and uneven. For example, in the case of EPBs, environmentalists in Taiwan seem to give high marks to those in Taipei, Kaohsiung, Ilan and Taoyuan. Occasionally, the news reports contain accounts of local EPB officials accused of accepting bribes or colluding with developers. This probably happens in every country.

A further complication is that much of the above structure is likely to change in the very near future. Starting in the late 1990s, both the KMT and DPP, responding to public pressure, began discussing ways to improve the process of environmental protection by consolidating agencies and improving the lines of authority. Ironically, this has been one of the few initiatives for which both of the two major parties have been in agreement. The overriding issue is a desire to re-organize the Executive Yuan, which has become bloated and inefficient, with more than 20 commissions or councils added to the originally authorized eight ministries and two commissions under the existing Organization Law. Within the hoped-for downsizing, on the environmental front the plan is to create a new Ministry of Environment and Resources (MER), which would bring together all central government agencies that deal with the environment, including the TEPA, and the various divisions of the COA, MOEA, Interior, Communications, as well as the Weather Bureau, Geological Survey, Bureau of Mining, and the Nuclear Control Commission. The MER also would include the National Land Use Planning division, and the current National Council for Sustainable Development in the Executive Yuan (see above).[45] In other words, the MER would be responsible for a huge portfolio overseeing environmental protection, water resources, mining, geological affairs, national land planning, national parks and scenic spots, meteorology, forestation, and regulation of nuclear power. There would be just four councils, in addition to 13 ministries: Council of National Development and Science and Technology, Oceanic Affairs Council, Council for Indigenous Peoples, and Council for Hakka Affairs. The latter two reveal much about the new status of Taiwan's two major minority groups, the aborigines and the Hakkas, who are enjoying a renaissance of identity and influence in contemporary Taiwan.

It is not certain yet if the TEPA head will become the first Minister of Environment and Resources. How to merge all these agencies and keep them satisfied will be a huge task. Potentially, if the TEPA is not the top agency within the MER, one could see the TEPA actually losing some power in such a structure. The MER will consist of three parts: Resource Use Planning, Pollution Control, and Nature Conservation. The creation of the MER will merge thousands of employees in many government bodies into one super ministry, the largest in the Taiwan government.[46] Supporters of the MER see two principal advantages from it: (1) reduction in size of government, by merging of agencies and (theoretically) reducing manpower through retirement and attrition; and (2) strengthening the environmental protection ability of the government, especially through effective national land use planning that should make environmental protection easier. It all looks good on paper; one can only hope that reality (after passage) matches anticipation. Hardly surprising, environmentalists in Taiwan are strongly for the MER in principle. The Legislature was scheduled to vote on this bill in 2006. However, frequent changes in the Premiership and other instability in the DPP government seemed to indicate that there could be further delays in bringing the proposal to a vote. As of early 2007 it appeared fairly certain that the political gridlock gripping the government would prevent bringing this measure to a vote in the legislature any time before the 2008 presidential election. That means MER will not become reality until at least 2009 or 2010.

Originally it was also proposed to create a Ministry of Ocean Affairs, which would include, among others, the existing Bureau of Coastal Zone Management in the Construction and Planning Administration. This proposal was highly controversial, given the strongly conflicting views within Taiwan as to how to develop and manage coastal zone resources, not to mention the various fishing zones and their problems (see Chapter 2). Now, it appears that compromise has reduced ocean affairs to a council, which will handle policy planning, promotion, and coordination of international oceanic affairs (such as deep-sea fishing agreements); oceanic science and technology; cultivation of talent in oceanic affairs; and coastal management.

Environmental Protection Administration[47]

Evolution of the TEPA

Since its inception, the TEPA has had a huge responsibility and an ambitious agenda. The success, or not, of the TEPA lies in part in the effectiveness of its top leader. As one of the key environmentalists in Taiwan put it, the head of the TEPA must meet several requirements in order to succeed: (1) he must have a clear strategic plan and know how to implement it; (2) he must have the '3Cs' skills (communication, coordination, cooperation), and the '3Es' skills (environmental, economic, engineering), in order to develop optimum strategies; (3) his image must be positive and he must have good political sense to win the trust of the public; and (4) he must set goals, and know how to integrate environmental protection with economic development.[48] The perfect leader does not exist, of course, and the persons who have successively led the TEPA over the years since 1987 have each had their strengths and weaknesses. Mostly Taiwanese, the heads have also been highly educated (all but two with Ph.D.s in the sciences or engineering, mostly from American and Japanese universities). They all have had considerable experience in environmental issues as administrators or faculty members in universities before taking on leadership of the TEPA.[49] Most of them have done their best to promote the policies of the TEPA and improve the environment, working against severe odds sometimes, in one of the toughest administrative jobs in the central government. In other words, the TEPA head often has to walk a fine line, struggling to find a middle ground that will satisfy both the hard-core environmentalists and the pro-growth supporters.

Brief biographies of the eight successive TEPA heads to date reveal the strengths and weaknesses they brought to the job and the many forces and interest groups they have had to contend with:

1 *Chien Yu-hsin* (1987–91)
Chien (a Taiwanese Hoklo) pushed very hard for environmental education and focusing on young people, on the rationale that for the long-run it is necessary to change peoples' attitudes toward the environment and the best time to start is when people are young. Chien is warmly remembered today

by environmentalists for his devotion to environmental education. Overall, he was relatively effective and got the TEPA off to a good start.

2 *Chao Shao-kang* (1991–2)
Chao (a Mainlander) was a former legislator in the Legislative Yuan when picked to run the TEPA. The mass media were strongly for him. He made some major decisions during his brief tenure, including closing of polluting factories, which led to some conflict with the central government and the KMT. He resigned after a short time in office because he wanted to enforce restrictive environmental policies, but the Lee Teng-hui government would not support him.

3 *Chang Lung-shen* (1992–5)
Chang (a Taiwanese Hakka) was formerly the first head of the Construction and Planning Administration (CPA) in the Ministry of Interior, which is responsible today for the Department of National Parks, which Chang established during his tenure at CPA. He was also former Deputy Minister of Interior, and Deputy Director of the Council for Economic Planning and Development. Chang is remembered best, thus, for his work in nature conservation and the excellent system of national parks that grace Taiwan today. Unfortunately, Chang could not work well with the media, and had a falling out with the Lee Teng-hui government over use of the coastal zone. Chang had the first thorough survey of the coastal zone carried out under his tenure, and established 12 coastal zone protected areas. His opposition to coastal industrial parks and tidal land reclamation proved too much though, and he resigned, evidence again of the toughness of the job of running the TEPA.

4 *Tsai Hsung-hsiung* (1996–2000)
Tsai (a Taiwanese Hoklo) was formerly Director of Urban and Housing Development in the Council for Economic Planning and Development (CEPD) in the Executive Yuan. Tsai is remembered by environmentalists as perhaps the least effective TEPA leader so far. He tended to compromise with big business and with the KMT central government too much, and hence the programmes of the TEPA did not advance much during his time at the helm.

5 *Lin Jun-yi (Edgar)* (2000–1)
Lin (a Taiwanese Hoklo) was probably the most controversial TEPA head. A professor of biology at Tunghai University, he was regarded as the 'father' of the anti-nuclear movement. Ironically, he became head just as the fourth nuclear power plant became a hot issue again, and was appointed by President Chen to spearhead the DPP's environmental agenda. But Lin proved too hot even for the DPP. Ostensibly, he was removed from office because he was in the US at the time in January 2001 when a Greek-registered (but Norwegian-owned) oil tanker, the *Amorgos*, grounded near Kenting National Park in southern Taiwan, creating the worst oil spill in Taiwan's history, and Lin supposedly did not respond to the problem fast enough. The underlying reason, however, is widely held to be Lin's opposi-

tion to nuclear power and politically incorrect public statements that he made against the KMT. Morale inside the TEPA was also said to be very low during his tenure, with rampant politics and division heads ignoring Lin. Thus, President Chen felt he had no choice but to replace Lin.

6 *Hau Lung-bin* (March 2001–October 2003)
Hau (a Mainlander), was son of the ex-premier and general, Hau Pei-tsun, a former powerful figure in KMT circles. Hau Lung-bin is typical of Mainlanders born on Taiwan into privileged circumstances, who received the best education in Taiwan and the US, and who see Taiwan as their home. Not uncommon with key administrators in the ROC government, such as former President Lee Teng-hui, Hau first worked as a university professor (1984–96) at National Taiwan University, the top school on the island, then joined the KMT and served in the Legislative Yuan (1996–9). Then Hau split from the KMT and helped found the New Party (see page 85). President Chen, seeking to form a consensus government that reached across the political spectrum and drew upon talent in both Mainlanders and Taiwanese, asked Hau to serve as TEPA head after Lin (see above) was forced to resign. Hau successfully led the quick cleanup of the oil spill at Kenting, averting a potential catastrophe.[50] Trained in agricultural chemistry and food science, Hau earned high marks from many environmentalists and was liked by many Taiwanese, although criticized by some for his support of nuclear power and incinerators for solid waste. One of his most successful policies was to enact the programme for eliminating the use of plastic bags. He seemed to possess nearly all the qualities needed to head the TEPA, as noted above, and should have had a long and successful run in his post.[51]

However, Hau was undone by the stress of the conflicting pressures that inevitably converge on the office of the TEPA head, and thus his case is very illustrative. Hau resigned because of a dispute over whether results of a referendum should overrule the decisions of environmental specialists. Specifically, residents of Pinglin township in Taipei County voted to approve building an access road to the Taipei–Ilan freeway, then under construction, even though the EIA had recommended against the project. Hau insisted on banning the access road, arguing that it would lead to a rapid increase in tourists who would pollute the catchment, which supplies the Feitsui Reservoir with water for the greater Taipei area. In other words, the convenience of tourists was felt to be less important than the greater good of the residents of Taipei as a whole. The broader issue was whether citizen referendums should be allowed to overturn environmental impact assessments. Lawmakers blasted Hau for overturning the referendum. The government did not completely back Hau, pointing out that EIAs are only 'advisory' and neither legally binding nor representing a final policy decision on a particular project. Moreover, environmental expertise has to be balanced with democracy and civic rights, the Premier said. The government seemed to be leaning in support of the citizen referendum and against Hau. Thus, Hau felt he had to resign, as a matter of principle.[52]

The Taiwan government thus has yet to come to grips with some fundamental questions: How much authority should the TEPA have? What role should EIAs play in construction projects? Should citizen referendums be allowed on development projects for which EIAs have been completed?[53] Tough questions with no easy answers.

7 *Chang Juu-en* (2003–5)

Chang (a Taiwanese Hoklo), who had served as Hau Lung-bin's deputy since April, 2001, was chosen by President Chen to replace Hau in October, 2003. Chang did his academic degrees in civil engineering at National Cheng Kung University in Taiwan, and then his Ph.D. at Tohoku University, Japan, illustrative thus of the other camp of key administrators whose academic links are stronger with Japan rather than the US. Chang had a distinguished record in environmental engineering and protection in various positions, including serving as Associate Dean of the College of Engineering at National Cheng Kung University (NCKU).

Chang got along well with the DPP, and followed Hau Lung-bin's policies very carefully. However, Chang turned out to be less forceful than Hau, behaving more like a traditional civil servant, a technocrat. Chang unexpectedly resigned in April, 2005, claiming that his position at NCKU would be lost if he did not return then (he was on leave of absence from his professorship). The insider story was that the Premier preferred someone else. He saw Chang as not 'political' enough (Chang was not a member of any political party).

8 *Chang Kuo-lung* (2005–7)

Chang Kuo-lung (a Taiwanese Hoklo), a veteran environmental protection activist nicknamed Taiwan's 'anti-nuclear godfather', was selected to succeed Chang Juu-en. Formerly a political vice minister in the Ministry of Examination and a Professor of Physics at National Taiwan University, Chang was educated at Yale. He had long been an anti-nuclear activist in Taiwan. Hence, his accession to the top environmental job in Taiwan was hailed by environmental protection groups. Chang was trying to tackle the problem of battery waste and the dioxin problem, and ran intro criticism in the legislature for that. He also relied very heavily on environmental NPOs, especially the TEPU, for persons he could appoint to the EIA Commission. There may have been a problem in this, because these appointees sometimes have difficulty working well in the TEPA and with government officials as well as business people. In other words, the TEPA may have been tilting too far to the left under Chang. There definitely were far fewer EIA approvals under Chang, compared to previous heads. As a result, some controversial issues led to disagreements between the TEPA and the Central Government (albeit led by Chen and the DPP), which feels the need to balance environmental protection with economic concerns, as noted earlier. Chang resigned, thus, in early 2007, another casualty of the intense pressures at the top of the TEPA, and was temporarily replaced by a DPP legislator while a search was undertaken for a new head of the TEPA (as of mid-2007).

Another leading environmentalist in Taiwan had some pessimistic thoughts about the relationship between the TEPA head and the various groups that the head must deal with in Taiwan. He pointed out that unless the Premier and President of the government are strongly for the environment, the TEPA head does not have much power. Legislators seldom look at the long-term welfare of the environment, as they are under constant pressure from industrialists and developers. Moreover, the public has only a very superficial sense of environmental consciousness, typically confined to issues that directly impact them individually, such as water and air pollution, trash, etc. The truly concerned and knowledgeable environmentalists able to look at the big picture are a tiny circle in Taiwan society.[54] Is this the real state of affairs, or just the opinion of a somewhat jaded activist environmentalist? Regardless, there is no question that the TEPA has one of the most challenging responsibilities of any governmental agency today. That challenge, and the TEPA's response to it, can be seen in the various branches and sub-units set up by the TEPA since its inception in 1987, and the various policies and programmes enacted in the different sectors of the environment.

Organization of the TEPA

The TEPA has undergone many changes since its beginning in 1987, of course, as this or that unit was added or dropped, units merged, responsibilities shifted, names changed, etc.[55] Although closely modelled after the EPA in the US and Japan, the TEPA has found over the years what organizational structure works and what does not in the unique political, economic, and social environment of Taiwan. Currently, thus, the TEPA classifies its *responsibilities* into two main groups, as follows:

- *Pollution control*, which consists of eight categories:
  water pollution control; air pollution control; noise and vibration control; soil and groundwater web; solid waste management; toxic substance management; environmental sanitation management; non-ionization radiation control.
- *Environmental management*, which consists of eight categories:
  environmental protection plan; environmental impact assessment; drinking water management; environmental quality monitoring; environmental analysis; environmental training; environmental education; environmental statistics.

The actual carrying out of these responsibilities, however, is done under some 15 divisions or branches of the TEPA, namely: eight departments, and seven other units of various functions (Figure 3.1). The activities, policies, and major accomplishments of each of these would require far too much space to document here. The full details are available on the TEPA website, and the government's annual update of its *Taiwan Yearbook* also provides a useful summary of progress on the environmental front. Following is a selective sampling of some of the developments and accomplishments of some key divisions of the TEPA.

Comprehensive planning

The Department of Comprehensive Planning in the TEPA is responsible for enacting environmental policy, promoting environmental education, and reviewing environmental impact assessments. In other words, this unit more or less oversees the broader mission of the TEPA as a whole. It monitors the progress of the TEPA across the board, and the administrators of this department particularly point to successes in seven areas of accomplishment by the TEPA as a whole:

1 Reduced pollution in air quality, water quality, drinking water management, toxic chemical management, garbage processing, industrial waste management, and resource recycling.
2 Establishment of the National Council of Sustainable Development (NCSD) under the Executive Yuan, to promote the long-range goal of sustainable development (see above).
3 Enactment of many environmental protection laws.
4 Building of many environmental facilities, such as sanitation, landfill, air and water-monitoring stations.
5 Enforcement of a formal environmental education curriculum and non-formal environmental NGO development.
6 Upgrading of environmental staff quality. TEPA personnel tend to be younger and more highly educated than is true of most other governmental organizations.
7 Involvement in international environmental protection affairs and organizations, as part of the global environmental protection network.

Basic Environment Act

One of the most important pieces of legislation handled by this department has been the Basic Environment Act. This landmark legislation is a broad statement of principles, something like a constitution on the environment, stating the fundamental ideals on which the people of Taiwan are in agreement. Understandably, with so many interest groups involved, passing this legislation was not easy. The bill had been under consideration since first proposed by the TEPA in 1988.[56] The most important feature of the law is that it makes environmental protection a prime consideration of economic growth, a stunning contrast with the early KMT days in Taiwan. In 2002, the Legislative Yuan passed the third reading of this law, meaning that it had been rejected twice before and had to go back down the ladder for revisions and further consultations with various groups. One controversial part of the Basic Law is to create a nuclear-free Taiwan, in line with the DPP's long-standing opposition to nuclear energy. The law also has a number of other important provisions, among which are:[57]

1 All levels of government will be required to protect resources, curb the emission of greenhouse gasses, and refrain from polluting the water table.
2 The central government should set up a Ministry of Environmental Resources to tackle pollution problems, and formulate a compensation system and emergency measures to handle future contaminations that might occur.
3 Citizens or NPOs will be allowed to file lawsuits against the government in cases where the state is believed to have been negligent in preserving the environment and conserving natural resources.
4 Environmental protection groups will be encouraged to form, and incentives will be provided to civil groups and businesses that engage in environmental protection.

In short, these are bold, strongly pro-environment measures that could go a long way in moving Taiwan toward the cherished goal of a 'Green Silicon Island'.

While the law was wending its tortuous way through the legislative process, the TEPA did not stand still. Instead, since its founding, the TEPA has promulgated more than 14 sets of laws (and amended them many times), covering a broad range of pollution and environmental issues: the Statutes establishing the TEPA; the Environmental Impact Assessment Act; the Air Pollution Control Act; the Noise Pollution Control Act; the Water Pollution Control Act; the Marine Pollution Control Act; the Waste Disposal Act; the Soil and Groundwater Pollution Remediation Act; the Toxic Chemical Substances Control Act; the Drinking Water Management Statutes; the Environmental Agent Management Act; the Environmental Dispute Settlement Act; Statutes Concerning EPA Inspection Organizations; and Statutes Concerning the Training of Environmental Personnel. Within these sets of laws, the TEPA has also put into effect dozens of regulations and executive orders, many of which have been amended and some rescinded. In other words, the TEPA was able to get action going on specific fronts in the environmental war, even though it was difficult to get agreement on the broader principles of the Basic Act until that act was finally passed in December 2002, with its 41 articles: 14 under General Principles; four under Planning and Protection; 16 under Control and Relief; five under Guidance, Monitoring, Incentives and Penalties; and two under Supplementary Provisions.

There are important differences, however, between 'laws' and 'regulations'. Laws must be passed by the Legislative Yuan, and have methods for punishing violators. As such, laws are more difficult to get approved. By contrast, regulations can be passed solely by the TEPA itself. Regulations are basically details pertaining to larger laws and are written according to the law. If one does not follow regulations, then one will be punished according to the provisions of the law.[58]

Three-year environmental action plan

Starting in 2005, the TEPA announced a Three-Year Action Plan, which integrated the efforts of all the major TEPA departments into one combined force for a span of three years. The purpose was to firmly put the provisions of the Basic Environmental Act into operation, adopting that Act's spirit of sustainability. Seen by some as a way for the then new EPA head, Chang Juu-en, to make his mark, the Action Plan contained six sub-plans:

1 The Model Environmental Lifestyles Plan, to employ a variety of rewards to encourage citizens to live environmentally friendly lifestyles, through such aspects as recycling, cleaning up household areas and local environs, developing community environment reconstruction plans and sustainable homes, promoting green consumer plans, strengthening environmental sanitation, reinforcing drinking water management, and others.
2 The Open Information and Full Citizen Participation Plan, to make environmental information accessible to all and encourage full citizen participation in environmental protection policies, through various forums and the Internet.
3 The Environmental Pollutant Reduction Plan, to specifically target dioxin, volatile organic compounds, heavy metals, toxic chemical substances, river basin pollution, and soil and groundwater pollution, with specific goals for each of these categories.
4 The Complete Sorting of Garbage for Zero Waste Plan, designed to ensure that resources are efficiently recycled and reused, to eventually achieve the goals of complete sorting and zero waste.
5 The Industrial Waste Control and Zero Waste Strategy, aimed at accomplishing the same objectives for industrial waste as the previous plan would do for garbage.
6 The Environmental Fate Monitoring Action Plan, calling for the completion of one environmental law forensics case verification plan for one pollution source or pollution site per year, to stamp out illegal dumping.

As often happens with plans such as this, however, when the leader promoting such is no longer in office, the plan gets diluted, changed, slowed down, or outright abandoned, as the new leader tries to make his mark. Such was the fate of the Three-Year Action Plan, insofar as it was promulgated under Chang. However, portions of it still are being promoted as goals of the TEPA, albeit not in such a condensed time period.

Public education efforts

The TEPA has launched a great many public education programmes or campaigns over the years, all designed to increase public awareness of environmental issues and to get the public directly involved in environmental cleanup and protection, in a bottom-up approach. Since the very first head of the TEPA started this effort, it has been one of the most important activities of the TEPA and definitely having an impact. During Chien Yu-hsin's time, the effort was directed mainly at primary and secondary high schools. After Chien, the TEPA gave money to NPOs (including the TEPU) and to community colleges to carry out public education efforts. This has been a very effective bottom-up approach that has paid off well in changing public attitudes and behaviour toward the environment, and justifiably could be ranked as one of the TEPA's biggest successes to date.

As illustrative of the multitude of educational efforts carried out, here is a summary of some of the programmes just for 2002:

1 Carried out environmental work on campuses, in line with the Three-Year Executive Plan for Environmental Education in Schools.
2 Selected 17 college groups to carry out the Environmental Navigator Plan to implement the plan of environmental education for colleges.
3 Conducted four group activities regarding environmental education and sustainable development, training 200 teachers. Selected the ten best education models and 20 short lectures, and set up the website for environmental education.
4 Made the slogan for limiting the use of plastic bags and designed a Chinese calligraphy competition and professional activities via the web.
5 Selected the top performing school, teacher, and student in carrying out environmental protection work.
6 Organized lectures for schools to proceed with the training and practice of limiting the use of plastic bags and plastic eating utensils (including Styrofoam).
7 Finished the manual for teachers and student activities regarding limiting the use of plastic bags and utensils.
8 *Green Mark programme:* The 'Green Mark' programme got under way in August, 1992, when the TEPA started giving these green labels, showing the earth wrapped in a leaf, to products that met TEPA standards for being recyclable, low polluting, and resource saving.[59] Article 96 in The Government Procurement Law says that bidders who possess green marks, i.e. produce environmentally friendly products, should enjoy preference in government bidding. Other benefits are stipulated in relevant procurement codes. Similar to 'green seals' used in the US, the green marks in Taiwan are designed both to help encourage businesses to be more environmentally conscious and also to encourage the public to purchase these kinds of products. The paper industry led in adopting the green mark programme, as part of the drive for recycling (see below). Dozens of products now have these labels on them.[60] By 2002, applications for use of the Green Mark logo had increased four fold. Starting 1 January 2007, the government set a target of having 80 per cent of procurement made by government agencies to be 'green products', indicating major progress in pushing forward this important programme.[61]

As one more example of efforts to involve the public in the environmental movement, and make it a bottom-up strategy as much as possible, in 2006 the

TEPA launched a competition to seek new ideas from the public about how to conserve resources, recycle waste, and reduce environmental pollution. Contributors could send their ideas to the TEPA via their website, and the winners would receive NT$10,000 (about US$335) for each suggestion adopted. The TEPA promised also that the best ideas would be published in TEPA manuals used in public education efforts.[62]

Environmental impact assessment

Environmental impact assessments (EIA) have become a staple of environmental programmes around the world in recent decades. An EIA is designed to minimize, if not prevent, the negative impact of a project on the natural, social, economic, or cultural environment. Developed countries, such as the US and Japan, have had EIA laws since early in their national environmental protection programmes, dating back to at least the 1970s. The idea for an EIA Law (EIAL) began even before the TEPA was established. EIA Acts were passed by the Executive Yuan starting in 1984, each one lasting five years. Once Taiwan got on the environmental bandwagon in the 1980s, it proceeded to develop its own EIA Law, which was formally passed in December 1994. Modelled after the EIAL of the US, Taiwan's EIAL lays out in chapters and articles very precise rules and procedures as to how projects should be planned so that they meet national (Taiwan) environmental standards, how the projects should be reviewed, and the penalties for violations.[63]

Understandably, the law had a tortuous and controversial birthing, as it pitted industrial interests against residents and government. The deliberative body that makes the final decisions on approval, or not, of projects is the EIA Commission, which has 21 members (seven from government, 14 from academia, nominated by a search committee composed of government officials and academics).[64] The Commission has reviewed well over 800 projects since it came into being, but handles only projects of 10 ha or larger, which are forwarded to the Commission from relevant ministries in the government. Smaller projects are handled directly by the local EPBs. The Commission's review process normally takes up to 50 days (by law), but sometimes as long as 100 days or more for really big projects, such as major highways, petrochemical plants, golf courses, industrial parks, etc. (see below). In the opinion of an environmental expert closely involved in the process, the EIA system works quite well, although not without some controversy inevitably.[65] This was especially true in the early years, soon after passage of the law, because business concerns either thought they could get away with perfunctory observance of the law, or were simply ignorant or negligent in carrying out the required steps and paperwork to get EIA approval for projects. Strict adherence to the EIAL often means added cost to the project. In the past at least, sometimes officials themselves were lax or easy going about enforcing all the rules and regulations of the EIAL.[66]

The Binnan Industrial Park project is a perfect example of the implementation of the EIAL and some of the problems and controversy that still revolve around this essential law in the TEPA arsenal. Binnan is a site in Tainan County, on the south-west coast of Taiwan, where industrialists proposed in the late 1990s to build a large industrial park. It would contain the seventh naptha cracker and a steel mill and, at more than 2,000 ha in area, the largest coastal zone development project in Taiwan's history. The Ministry of Economic Affairs, through its Industrial Development Bureau (IDB), which has often been viewed by environmentalists as very pro-business (especially in the past), submitted a draft EIA report to the TEPA in December 1997, on behalf of the project builders, Tuntex Petrochemicals, Inc. and the Yieh Loong Group. The TEPA, through the EIA Commission, found the EIA report inadequate and sent it back for revision. The concerns of the TEPA, and outside environmentalists also, centred around a variety of issues: the impact on the natural ecology of the lagoon where the project would be built, carbon dioxide emission control, limits on water consumption (the site is in a drought-prone part of Taiwan and would require desalinization of seawater by the project), and protection of the migratory black-faced spoonbills that over-winter in the area. The TEPA laid out dozens of specific requirements on all of the above in order for the project to be approved. The EIA passed shortly before Chen became President in 2000. However, the project still had not started by 2007 because of a scandal in the Tuntex Corp. The speculation is that this project will never get built.

Looking at the review document from the perspective of the project builders, it is easy to see why industrialists are sometimes discomfited by the EIA process. The Commission's requirements (not just recommendations) undoubtedly will add significant costs to the project in both money and time. The TEPA offers the rejoinder that this is the necessary cost of large and potentially very damaging projects like that at Binnan, and that such projects must build into their cost projections the appropriate environmental protections mandated by the TEPA, and then pass those costs on to the consumer. It is better to let the consumer pay for the protection than for society to suffer the environmental consequences of unregulated development. This fundamental reality is sometimes difficult, though, for consumers as well as developers to swallow. It is a learning process.

Water quality protection

There has been progress in improving Taiwan's water resources, through the TEPA's Department of Water Quality Protection and other agencies, but still much remains to be done. Of Taiwan's 118 rivers, 24 are under the central government and 94 under local government. Just 24 of these rivers provide 85 per cent of Taiwan's potable water supply. The degradation of river water quality peaked in 1996, at a ratio of 15 per cent; since then, the proportion has steadily diminished, and the goal for 2011 is just 8 per cent. To monitor water quality, some 311 sampling stations had been established throughout the island as of 2003, including 38 in Taipei's badly polluted Tanshui River basin. The TEPA had a priority project running between 2002 and 2007 that focuses on 13 rivers of high priority for clean-up.

The main pollutants of rivers are domestic sewage and industrial waste, coming mainly from urban communities. The rate of progress on the sewage front for Taiwan as a whole can be seen in Table 3.1, while the same categories are revealed in Table 3.2 for the Taipei Metro Area. Progress is being made, but slowly. By 2005, just 38 per cent of Taiwan's population had access to some form of sewage treatment, while only 13 per cent were connected to piped sewer systems. The figures for the Taipei metro area were not that much greater, with just over half the urban population having access to some form, and about a quarter to piped sewage systems, remarkably low figures for one of Asia's supposedly most modern and sophisticated cities. The projections to 2010 show continued improvement, to be sure, but still it will be many decades, in all likelihood, before the sewage treatment figures for Taiwan (and Taipei) match those in highly developed countries. Many factors account for this slowness, including politics and the high density of the built-up areas and hence high costs of installing sewer lines and treatment plants. Nonetheless, this deserves to be a higher priority for Taiwan's environmental efforts.[67]

As for industrial wastewater, the TEPA has the daunting task of trying to monitor more than 17,000 factories (out of an even larger total number) around the island. Sampling and examining the effluent quality at the outfalls of factories is the primary means of carrying out this monitoring. This approach, though,

Table 3.1 Domestic sewage control in Taiwan

Goals	*Year 2000*	*Population served*	*Year 2005*	*Population served*	*Year 2010*	*Population served*
Pop. ('000)	22,240		23,120		23,910	
Domestic sewage connected		1,120 (5%)		3,080 (13%)		5,370 (22%)
River run-off intercepted		880 (4%)		1,440 (6%)		1,850 (8%)
Community sewer systems		1,100 (5%)		1,680 (7%)		2,380 (10%)
On-site treatment facilities		2 (0%)		18 (0%)		44 (0%)
Treatment facilities in new buildings		260 (1%)		1,120 (5%)		2,160 (9%)
Septic tanks clean-up		280 (1%)		450 (2%)		650 (3%)
Sewage reduction[1]		480 (2%)		960 (4%)		1,680 (7%)
Total served		4,122 (18%)		8,748 (38%)		14,134 (59%)

Source: TEPA Yearbook 2004.

Note
1 Domestic sewage reduction due to educational persuasion.

Table 3.2 Domestic sewage control in Taipei metro area

Goals	*Year 2000*	*Population served ('000)*	*Year 2005*	*Population served ('000)*	*Year 2010*	*Population served ('000)*
Metro area pop. ('000)	6,350		6,430		6,510	
Domestic sewage connected		1,006 (15.8%)		1,522 (24%)		1,938 (30%)
River run-off intercepted		757 (12%)		1,060 (16.5%)		1,020 (15.7%)
Community sewer systems		317 (5%)		400 (6%)		470 (7%)
On-site treatment facilities		0.8 (0%)		0.9 (0%)		1 (0%)
Treatment facilities in new buildings		9 (0.1%)		22 (0.3%)		27 (0.4%)
Reg. septic tank clean-up		90 (1.4%)		352 (5.5%)		443 (6.8%)
No reg. septic tank clean-up		41 (0.6%)		30 (0.5%)		24 (0.4%)
Sewage reduction[1]		60 (1%)		120 (2%)		190 (3%)
Total served		2,281 (36%)		3,507 (55%)		4,113 (63%)

Source: TEPA Yearbook 2004.

Note

1 Domestic sewage reduction due to educational persuasion.

does not prevent or detect illegal activities, such as bypass piping and dilution of wastewater. As already observed in Chapter 2, illegal activities and improper disposal of wastewater still regularly occur. This problem will take decades to resolve.

The pig-raising industry is part of the problem. With about seven million pigs still, the pollution generated by them is likened by the TEPA to equivalent to an additional 30 million people. Taiwan thus is trying to downsize the industry and its pollution, helped by Taiwan's admission to the WTO and hence the need to reduce the number of pig farms in order to maintain competitiveness with imported pork. The gravy days for pig raising are over. The TEPA uses subsidies to pig farms for installing pollution treatment facilities and for voluntarily downsizing the number of swine, especially targeting pig farms located in water conservation areas where they are most prone to polluting nearby rivers and streams.

One relative success has been in the greater Kaohsiung area, where the city's new water system came into operation in November 2003. This was a huge project that involved integrating new water sourcing facilities with three advanced water purification plants, and relocating more than 400,000 pigs being raised along the Kaoping River. The more than 2.5 million residents of Kaohsiung no longer have to buy and haul drinking water home. In addition, clean-up over a period of a number of years of the Love River, which flows right through downtown Kaohsiung, has vastly improved what was once virtually an open sewer. The two banks of the river have been redeveloped, as a result, with various private and public facilities that provide a handsome year-round recreation site for residents and visitors, helping enormously to enhance the public image of Kaohsiung and the quality of life there.[68]

By contrast, the Tanshui River in the Taipei Basin remains a major problem, even though the proportion that is still 'seriously polluted' has fallen officially to 13 per cent. The government thus recently announced a three-year, NT$1 billion (US$33 million) programme to further improve the Tanshui, including seven wastewater processing centres and three ecological engineering projects.[69] In 2005, the Executive Yuan proposed an eight-year, NT$80 billion (US$2.6 billion) programme of public works in water conservation and flood prevention to repair Taiwan's damaged rivers and coast, in a kind of crash programme that was hoped to accelerate progress in this notoriously slow endeavour. The programme would launch projects in the most urgently needed areas of 22 rivers, 50 water catchment systems and 51 km of coastal dikes.[70] The government pointed to the Keelung River project as an example of what could be accomplished elsewhere. That programme, begun in the early 1990s, cost billions of Taiwan dollars, but eventually tamed the river, especially because of the Yuanshanzih flood diversion tunnel (sluiceway), which helps redirect run-off from the upper reaches of the river to the sea on the north-east coast, thus saving the often flooded townships of Rueifang and Hsichih (Sijhih)[71] when future typhoons and heavy rains hit the area. The Keelung project was highly controversial when proposed, however, just as has happened with the new eight-year project, which is on an even grander scale.[72]

In line with the above, Taiwan's 57 water reservoirs are carefully monitored, and those needing improvement are worked on. The TEPA uses the Carlson Index, which measures transparency and levels of chlorophyll and phosphate.[73] Since 1997, the number of reservoirs considered suffering from eutrophication and hence deficient in water quality fell from 15 to just eight in 2006. The Sun Moon Lake Reservoir in central Taiwan, built originally by the Japanese, has the best water quality. By contrast, the Shihmen Reservoir in Taoyuan County is one of the worst (along with Fengshan in Kaohsiung County, the Mingte and Liyutan Reservoirs, which supply Miaoli County and the greater Taichung area, Taipei's Feitsui Reservoir, Kaohsiung's Chengching Lake, and Jingmian in Tainan County).[74] Built some 40 years ago, and now substandard, the Shihmen Reservoir is among the most turbid, especially after heavy rains associated with typhoons. Such conditions created a local crisis in August 2005, with temporary but severe water shortage.[75] A vast amount of silt has accumulated behind the Shihmen Dam because of deforestation and degradation of slopelands in the watershed, a problem not solved by yearly flushing of silt in the dry season.[76] The task for the TEPA is huge, again because of high population density and the difficulties in regulating or eliminating (often long-established) polluting activities within the catchment areas of reservoirs, such as pig raising, landfill sites (dumping sites for construction waste or garbage), improper cultivation and use of agricultural chemicals, illegal construction, etc. Much of this falls under the TEPA's programme to stop what it calls 'non-point sources pollution'.

Water for Taiwan's people is also obtained, of course, from groundwater sources, which currently provide potable water to about 15 per cent of the population. By 2006, the TEPA hoped to have a total of some 430 monitoring wells constructed around the island to monitor groundwater quality. All of the efforts noted above are aimed, of course, at stopping further pollution of the groundwater, reservoirs, rivers and streams. The issue boils down to Taiwan's basic water balance. Annual rainfall has averaged 75 billion tons in recent years, with four billion tons seeping into the water table. However, consumption of groundwater, especially in the central and southern coastal areas, averages 6–7 billion tons a year, resulting in chronic deficit. Thus, rainfall and run-off catchment becomes a top priority, as does efficiency in use of water.[77]

Marine pollution is a moderate problem. Since passage of the Marine Pollution Control Act in 2000, the TEPA has established 97 ocean water quality sampling stations. In general, the coastal waters are in fairly good condition, with occasional signs though of organic pollution from animal husbandry in the south, and coliform concentrations in the north due to the high population density there.[78] Since the near-catastrophe with the Amorgos oil spill in 2001 off southern Taiwan (see previous section), the government has perfected its procedures and ability to deal with coastal oil spills.

A persistent problem with water in Taiwan is not with its quality but quantity, and how water is used. According to the Environmental Quality Protection Foundation, Taiwan ranked 40th among 148 countries worldwide in 2003 in terms of the Water Poverty Index (WPI), an index developed in the UK that uses five indices (water resources, access, capacity, use, and environment) to assess the impact of water scarcity on public welfare. Taiwan's score put the island in about the same rank as the US, Germany, and Japan. However, Taiwan ranked 141 in use of water, indicating serious wastage. The average Taiwanese uses 350 litres of water per day, compared with 150 litres in the US or Europe. The primary reason is the cheapness of water in Taiwan, averaging just NT$9 per unit for the whole island, and NT$7 in Taipei, versus NT$40 in the US and Europe. Thus, people in Taiwan tend to waste water.[79]

Air quality protection

While improving, the air quality in Taiwan still is not ideal, and perhaps even more importantly still below peoples' expectations. A TEPA poll conducted in the spring of 2004 found that a majority of the respondents felt there had been no improvement at all in air quality between April 2003 and April 2004. Those people also felt that emissions from motor vehicles, factories, kitchens, and construction sites were most in need of curbing.[80] Polls like this reveal two things: (1) people base such opinions on perceptions, not scientific measurement; and (2) public perception is half the battle in environmental protection. Public polls aside, the air in Taipei and Kaohsiung is in fact better than it used to be. As examples of improvement: sulphur dioxide decreased from 0.27 ppm in 1990, down to just 0.003 in 2003; the PSI exceeded 100 on only 2.7 per cent of the days in 2003 (compared to 16.3 back in 1990) (Tables 2.1 and 2.2). In the heavy industrial district of Kaohsiung/Pingtung, the PSI was above 100 on just 6.6 per cent of the days in 2003 (vs. 18.4 in 1994). The goal anticipated for 2006 was to be 2 per cent, and a mere 1.5 per cent by 2011.

The initial Air Pollution Control Act was enacted back in 1975, even before the TEPA existed, but it has been revised and greatly strengthened since then, and now empowers the government at all levels to establish air quality standards for eight different areas on the island. An Air Quality Monitoring Network has been established, with 74 stationary automatic air quality monitoring stations, two mobile stations, and an air quality laboratory. In addition, there are eight photochemical monitoring stations. Hence, measurement of pollutants is the first and very important step in air quality protection. The current levels of those pollutants were noted in Chapter 2.

Air quality improvement measures include stringent emission standards for industrial plants and motor vehicles, promotion of low-pollution transportation vehicles, strict standards on composition of petroleum products, increased inspection of construction sites, and road cleaning. The TEPA also levies an Air Pollution Control (APC) fee on stationary sources (factories, construction sites) and mobile sources (motor vehicles). This fee covers suspended particulates, nitrogen oxides, sulphur oxides, and hydrocarbons.

The APC fee, which has been collected since 1995, generates close to US$200 million in revenue each year for air pollution control programmes, including inspections, enforcement measurements, promotion of clean energy resources, subsidies and incentives for reduction of pollutants, and international environmental efforts in air pollution.[81] The APC is a highly effective way of charging polluters and providing them with financial incentives to reduce emissions. To avoid public perception of the fee as a form of tax (although it really is just that, in another guise), the APC is administered by a separate foundation, which also decides how the collected fees will be allocated and used. Thus, a complex menu of fees, on a sliding scale, has evolved. As just one example, a charge is applied to coal, fuel oil, petroleum coke, diesel and leaded gasoline, but liquefied natural gas or liquefied petroleum gas are exempted. The APC is a good example of what environmentalists internationally have long argued for: make polluters pay for their pollution, rather than just use government funds to clean up after the fact.

The TEPA recently announced a new initiative to cut auto emission levels even further in the three key cities of Taipei, Taichung, and Kaohsiung. The plan targets the years 2006–8 with a goal of reducing air pollution an additional 30 per cent. This will be done by designating more bicycle lanes,[82] increasing production of liquefied petroleum cars to 18,000, adding 12 more stations for natural gas refuelling, promoting the use of hybrid cars, and the eventual elimination of all two-cycle engine motorcycles.[83] Since 1997, all motorbike owners have been required to have an annual inspection, and some 1,900 inspection points around Taiwan conduct spot checks. Owners discovered not to have carried out the inspection are fined NT$3,000 on the spot. Owners are entitled to one free annual emissions inspection. TEPA data show that the inspection system had achieved a rate of 65 per cent of all motorbikes by 2002. Taiwan's emission standards for motorcycles are now among the most stringent in the world.

Solid waste management

With each person in Taiwan generating an average of over one kilogram of trash per day, or some 8,400 tons per year, and with trash such a visible part of environmental degradation (see Chapter 2), the TEPA has viewed this problem as one of its top priorities. The goal, overall, has been to improve the process of disposal, promoting incineration over landfills, and encouraging recycling. To put it another way, the government urges citizens to 'reuse, reduce, recycle'. The results have been quite good. In many ways, Taiwan now has one of the world's most rigorous solid waste handling programmes, superior to that found in most US communities. Several strategies are employed in Taiwan.

First, since 2000, the government has required everyone to purchase official trash bags, the cost of which replaced the former garbage collection fees. Thus, each household pays a different rate for trash, depending on how many trash bags they put out on collection day. The more one throws away, the more one pays. This approach not only shifts the burden to consumers and encourages reducing trash, it also is more sanitary in that the new bags are much stronger and less likely to break and pollute the environment before being hauled off.

The second strategy is recycling.[84] Under the TEPA's Resource Recycling and Reuse Act of 1997, recycling actually began on a partial basis, with containers put along kerbsides by the government, but compliance was haphazard and inadequate. Compulsory garbage classification was implemented in January 2005 in ten cities and counties, as a test of implementing the programme island-wide. The test was so successful, the government made the programme mandatory for everywhere in Taiwan starting 1 January 2006. Under the initiative, citizens are required to sort their trash into three categories: ordinary garbage, recyclable garbage, and food scraps (kitchen leftovers). Separate containers are provided for each. Recyclable items include common materials such as plastic bottles, paper, clothes, tires, furniture, batteries, lamps, balls, film negatives, raincoats, wooden toys, compact discs, etc. Garbage that is not properly separated is rejected, and the offenders fined between NT$1,200 to $6,000 (approximately US$48–$240), depending on the severity of the offence. Money seems to talk, as the TEPA found a 90 per cent compliance rate in the test areas in 2005, and is hoping for similar results island-wide eventually. Recycling significantly reduced the volume of trash in those test areas, as one would expect. The food waste is also recycled, with about 75 per cent being used as pig feed, and the remaining 25 per cent turned into fertilizer through composting. Houses in Taiwan do not employ garbage disposals, so all food waste has to be taken out in the trash collection process. Hence, food recycling makes very good sense, and has been highly successful. Moreover, food waste used to be one of the most environmentally polluting items, but now is being properly collected and reused in socially beneficial ways.[85]

The impact of recycling is evident from the data released by the TEPA. Taiwan's recycling rate rose from 6 per cent, or 554,000 tons, in 1998 to just over 20 per cent, or 1.56 million tons, of recycled waste by 2004, a considerable accomplishment in just six years. Conversely, the average daily volume of garbage collected island-wide dropped from over 24 million tons in 1998 to about 16 million tons in 2004, or a per capita decline from 1.35 kg to 0.71 kg in the same period. The goal was to boost the overall recycling rate to 25 per cent of all garbage by 2007.[86] That goal was actually reached in May 2006, with the recycling rate of over 26 per cent. Moreover, the country's total waste volume had been reduced by 25 per cent, and the average daily amount of trash per person declining to a low of just 0.618 kg.[87]

Yet, there is still much to be done. High in priority are batteries, waste paper, and Styrofoam, which still had recycling rates of only 17 per cent, 46 per cent, and 42 per cent, respectively, by late 2005. A promising sign was that over 90 per cent of the public polled in early 2006 was in support of the recycling policy, although there were still complaints about insufficient and too small trash cans in public places.[88] The gradual phasing out of plastic bags, and plastic and Styrofoam containers and utensils, is a big part of this programme, and making significant progress. The goal is to replace such items with biodegradable or reusable materials and containers. Taiwan is well ahead of much of the rest of Asia (not to mention the US) in this progressive stance on one of the most pervasive environmental problems worldwide.[89] If customers insist on using plastic bags, they are charged for them (NT$2, about US$0.06, per bag). Public support for this programme is strong, though, and the TEPA reports a noticeable reduction in the volume of trash as a result.[90] One report noted that by October 2005, the quantity of disposable utensils thrown away had been reduced by 87 per cent; for plastic bags specifically, the reduction was 58 per cent, while 77 per cent of those polled said they had 'cut back' on use of plastic bags, and an even greater number supported the ban on use of disposable eating utensils in school cafeterias and restaurants in government buildings.[91] Unfortunately, the TEPA was forced to back down in early 2006 and allow operators of Taiwan's innumerable portable food stalls to be exempt (at least temporarily) from the plastic bag ban.[92] In short, while success has been less than total in the fight against plastic bags and disposable containers/utensils, Taiwan has made major strides forward.

Ordinary garbage that is not food waste or otherwise recyclable has to go somewhere. Historically, as already noted in Chapter 2, this went into landfills. Since the TEPA began, the government has moved aggressively to phase out landfills and promote incineration as the most environmentally friendly way to dispose of this portion of trash. A total of 21 incinerators was originally planned for Taiwan, with an end to all landfills by 2020.[93] Modern incinerators, with their advanced technology and emissions control devices, are supposed to be pollution free and have been promoted to the Taiwan public as such. By the end of the 1990s, however, Taiwan was beginning to find two major problems with incinerators: emission of dioxin, and production of large amounts of ash. As a result, a groundswell of public opposition to incinerators began to emerge, forcing the TEPA to rethink its policies on incinerators and landfills. All large-scale incinerators are now required to have dioxin control design and equipment.

A total of 20 major municipal waste incinerators were in operation by mid-2006, producing over 775,000 metric tons of ash. In addition to the 20 in operation, one incinerator was finished but not operating, two were under construction, seven were in planning. The TEPA estimated that by 2007, a total of 26 incinerators should be operating and able to handle the estimated 22,000 tons of garbage generated daily in Taiwan.[94] Ash reuse technology is available, and the TEPA has hopes to implement that, and to encourage private incinerators, through financial incentives, to adopt such technology also. The ash is made into bricks and other construction materials, and theoretically can be 100 per cent reused, as already achieved in Japan. But now, anti-incinerator feeling puts the government in a dilemma.

A notorious instance of backlash against incinerators took place in Hsinchu in 2005–6, when residents living near the Hsinchu Science Park protested against the incinerator located in the park. That incinerator was built in 2003 and in operation since July 2004, as an industrial facility for drying, or burning, sludge discharged from the park's wastewater treatment plant, some 60 tons per day. However, it turned out that the incinerator had been constructed without an EIA, because of a loophole in the law, and without informing some 300,000 neighbouring residents and 110,000 park employees. Moreover, three times the incinerator was found to be emitting lethal levels of dioxin. Thus, the TEPA ordered the closure of the incinerator in January 2006, amidst much controversy and finger-pointing. Hence, this incident was another demonstration of the power of an NGO (the Hsinchu Science Park Anti-Incinerator Association) successfully marshalling public protests to achieve its goal.[95]

It was also a demonstration of bad implementation of a basically sound strategy for disposal of solid waste (i.e., incineration). Somewhat similar to the anti-nuclear movement, it is unclear how permanent and deeply based is the anti-incineration movement in Taiwan. Some people obviously do not trust the technology of incinerators, because of the uncertainty about emissions of dioxin. This is one of the primary reasons why the TEPA has been pushing recycling in recent years, in order to reduce to the absolute minimum the amount of solid waste that must be either put in landfills or incinerated. As of mid-2006 there were enough incinerators to handle the volume of garbage for Taiwan as a whole. The TEPA had not yet determined how many incinerators would be shut down, and was in the process of launching a 'Strategic Environmental Assessment' to determine a coherent waste disposal policy for the years ahead. Various ad hoc measures are already being implemented in places. For example, recycling has reduced the volume of garbage in Taipei City by one-third, so Keelung and Taipei are sharing facilities. Keelung sends some of its garbage to Taipei for incineration, while Taipei sends some of its garbage to Keelung for landfill. Some existing landfills accept remnants left over after incineration. The overall objective is to find the right mix of recycling, incineration, and landfills that will take care of all of Taiwan's garbage properly while simultaneously satisfying public concerns.

The TEPA began in 1996 to develop plans for restoration of landfills into public parks. Current restoration work consists of two stages. The first involves sealing the landfill surface, to stabilize the landfill and prevent secondary pollution. The second stage consists of the actual development of the park and recreation facilities on top. As of the end of 2002, a total of 180 landfills, covering some 486 hectares, had been restored. This restoration work is part of the TEPA's public relations effort to soften public opposition to landfills. One controversy swirls around the fact that new landfills on slopes are exempted from EIA if the area involved is smaller than two hectares. Some of these landfills are being used for toxic wastes, and in some cases the construction of these landfills involves local gangsters. The TEPA has countered that future landfills must be built by local governments, which must keep the public better informed.[96] Officially, there were 155 landfills in operation in early 2006, with 34 more in planning. However, given the public mood against landfills, it may well be that those additional landfills will not be used, as the government has a so-called 'zero waste' goal for the year 2020. That means, theoretically, that all trash will be recycled or incinerated by that date. The government is basically in a no-win situation, because nobody wants to be near either incinerators or landfills, of whatever type, in the proverbial

NIMBY (not in my back yard) syndrome. The public has to be educated about the benefits and costs of each method. That is a long process, still far from complete. Transparency in the planning and construction of such facilities is vital to maintaining public support.

Industrial and toxic waste control

As of 2003, Taiwan's thousands of factories generated nearly 13.5 million tons of waste, of which just over one million tons was hazardous. Since 2001, manufacturers must assume responsibility for managing their waste, with heavy fines and even prison sentences awaiting violators. Nearly 20,000 chemical substances are regularly used in Taiwan, of which some 6,000 are highly toxic. Of the latter, 252 are formally listed by the TEPA as requiring permission from the government for production, import, export, sale, or use. As noted in Chapter 2, it is these toxic substances that pose perhaps the greatest danger in Taiwan, partly because they have been improperly disposed of for decades around the island. So now the government has to clean up the illegal waste sites, while also carrying out ongoing monitoring of existing factories to prevent further illegal dumping and proper disposal of wastes, toxic or not.

An example of success in this difficult endeavour has occurred in Miaoli County, where illegal dumping apparently has been stopped by the first-ever joint inspection project between the central government and local authorities. Remote areas in the coastal townships of Chunan, Tsaochiao, and Houlung, because of their relative isolation, had become notorious sites for illegal dumping. In 2004 alone, 16 cases of illegal dumping of industrial waste were discovered, with trucks dumping the waste with seeming impunity right along the roadside. Starting in June 2005, the TEPA and Miaoli County government stationed inspectors in the area and, with assistance from local residents, caught a number of violators, who were fined (the maximum fine, however, was under US$10,000). This was enough to scare off other potential violators, and the problem has almost dried up there.[97] The most important lesson of this project was showing the value of collaboration between the public and government agencies at different levels. The TEPA clearly cannot do the task all alone.

In the case of polluted farmland, the TEPA claims a high rate of success. As of the end of 2002, farmland polluted by excessive heavy metals (mainly through the use of polluted irrigation water) totalled 261 hectares (183 in Changhua County, 33 hectares in Hsinchu City, and over 11 hectares in Taoyuan County). Supposedly, 93 per cent of that damaged land has since been cleaned of the toxic substances, and the land could be ready for replanting within another year or two. Crops grown on the polluted farmland were rooted out and destroyed, and the owners prohibited from growing anything further until approval from the TEPA. The farmers are being reimbursed in the mean time for loss of production.[98] Some environmentalists remain unconvinced, however, that the TEPA figures accurately reflect the true scale of the problem.

E-waste control

This problem was introduced in Chapter 2. The Waste Electronics and Electrical Equipment Directive (WEEE) went into effect in the EU in late 2005 and early 2006, with the goal of preventing creation of waste from disposal of electronic equipment. The Act makes producers responsible for paying for the recovery, recycling or disposal of their waste. The Restriction on Hazardous Substances (RoHS) Directive aims to stop hazardous substances from getting into landfills. This became effective in July 2006, and bans six substances from being used in electronic and electrical equipment, including the heavy metals cadmium, mercury, lead, and hexavalent chromium, as well as two flame retardants (PBDE and PBB). These new environmental restrictions on electronic goods adopted by the EU pose a real challenge to Taiwan, with its more than 30,000 manufacturers that export 44 types of electronic and electrical products to Europe each year. In 2004, Taiwan exported about US$8 billion of such goods to Europe, accounting for about 2.5 per cent of Taiwan's GDP. For Taiwan producers (and all others as well), this means they are responsible for: funding the pick-up, recovery, recycling and disposal of items from community collection points; labelling a product properly when they put it on the market and providing a guarantee to pay for its treatment/recycling at the end of its life; providing information on reuse and treatment of their product within a year of its going on the market.

The data for early 2006 showed 164 of the 972 companies in Taiwan that export EEE goods to Europe still had not prepared for the new rules, which went into effect on 1 July.[99] The Industrial Technology Research Institute in Hsinchu has been helping companies accomplish this since word of the Directives first appeared in 2003. The slowness of response by some Taiwan companies is believed to be mainly due to the relatively small size of most Taiwan companies. Compliance adds to the production cost. Even for large companies, such as Acer, the world's fifth-largest personal computer vendor and the top notebook brand in Europe, the company estimated the new rules would add about US$50 million to its operating costs in Europe. Taiwan's government basically is telling its manufacturers, large and small, to become so-called 'green suppliers', or they risk losing their competitive edge.[100]

Indeed, the new EU rules are certain to be gradually adopted in other countries, including the US, where individual states are already moving in that direction (e.g. California). The costs of this compliance will eventually be passed on to consumers, of course, and once all manufacturers are following these environmentally friendly rules, the negative impact on profits will disappear. But that process may take several years, and hence Taiwan's electronic manufacturers face perhaps a difficult decade ahead (not even taking into account other facets of this highly competitive global industry). The stakes for Taiwan are high. The EEE industry is the most important sector of the economy. In 2004, it had a total production value of about US$170 billion, or nearly 50 per cent of total industrial production on the island. EEE exports were worth some US$87 billion, about half of the island's total exports. Taiwan's survival depends on successfully adapting to this more restrictive marketing environment for EEE products. The Ministry of Economic Affairs (MOEA) launched a Green Project in 2005 to further help companies carry out the required changes.[101]

The winner, in the long run, will be the environment. Greenpeace International thinks the EEE industry needs to go farther and faster to produce completely non-toxic devices, and demonstrated at the Computex Taipei Technology Trade Fair in June 2006.[102] In the short run, though, the environment in poorer countries may well suffer, as electronic goods that fall outside the new rules are shunted off to those less restrictive markets. One can already see this, for example, in the immense volumes of discarded cell phones, much of which end up in places like Latin America.[103] Sadly, this practice of dumping discarded or unwanted electronics of all kinds is increasingly prevalent in poorer countries, or places like China that welcome such cast-offs in order to harvest raw materials from them but do not always dispose of the remnants properly. Ironically, China has become one of the biggest dumping grounds.[104]

The TEPA has come up with one approach to solving the problem of discarded batteries, part of the EEE waste dilemma. The idea is to require consumers to exchange used batteries when they buy new ones. If they have no used ones for the exchange, they would have to put down a deposit that would be used as a discount when they return for new batteries again. The main difficulty would be in getting the multitude of small retail outlets in Taiwan to comply with the programme.[105] Nonetheless, this idea further supports the principle, widely approved by environmentalists, of making the consumer pay for the environmental impact of products they wish to use. Actually, the TEPA, through its Bureau of Solid Waste Management, is exploring many options and types of programmes to solve the problem of waste batteries.[106]

The 'Ten Nightmares'

The political and economic forces buffeting the TEPA from all directions are nowhere more apparent than in the ongoing controversy about the so-called 'Ten Nightmares'. This highly charged name was attached by critics to ten major public works projects that the government has been proposing or is actually in the process of building. In 2004, the government launched a five-year, NT$500 billion (US$14.7 billion) ten New Major Construction Projects, continuing a tradition begun decades earlier under the KMT of pushing major infrastructure projects in packages like this, as a way of promoting economic growth while demonstrating the vigour of the government. Criticism of such projects was often difficult in the past; now it is routine. Of the total projected cost, 80 per cent will be borrowed, and the remainder is to come from government revenues.[107]

The critics come mainly from more than 30 very active environmental NGOs, such as the Taiwan Environmental Protection Union, Greenpeace, the Green Formosa Front, the No Nuke Taiwan Union, the Ecology Conservation Alliance, the Wilderness Society, and the Green Citizens' Action Alliance, which have been among the most vocal critics of the DPP and the Chen Shui-bian administration. They argue that Chen and the DPP have sold out the environment, in the interest of promoting economic growth at any cost. They cite, for example, what they see as inappropriate development projects having their EIAs approved by the government in recent years, undermining the claim by the Chen government that its long-range goal is for sustainable development. The critics also point to what they see as a great range of 'unsolved' environmental problems, including illegal waste dumping, unregulated dioxin levels in food, the lack of policy for treating radioactive waste, invasive tourism activities in ecologically sensitive areas, over-fishing, and others.[108]

These kinds of projects are reminiscent of pork barrel public works projects carried out in Japan over the years by LDP governments eager to court local voters, and as a stimulus for economic growth in times of recession. Indeed, such practices are not uncommon in many countries around the world, including the US. Too often, such projects are carried out without proper EIAs and with often negative impacts on the environment, or where there is no real proven economic need for the project and thus with great waste of public funds.

The ten projects and their current status are as follows:

1 *Fourth nuclear power plant*
As already noted in Chapter 2, this plant is under construction, after many years of delays and embroilment in politics. Long past the point of no return, the project will be completed, but then, if future governments remain true to current pronouncements by both the Pan Blue and Pan Green camps, this plant will eventually be closed down and deactivated along with the other three nuclear plants. The future of nuclear power in Taiwan, however, remains very uncertain, and current claims may not be upheld in the future, given the volatility of the political environment and the pressing demands for energy in Taiwan (see also Chapter 5).
2 *Suao–Hualien freeway*
A classic pork barrel project with an estimated price tag of NT$100 billion, this freeway would link Suao and Hualien on the East Coast via a multi-lane freeway (Highway 5) that would never be fully used, and with potentially bad environmental impact because of the mountainous, difficult terrain it would pass through (and hence require many tunnels and bridges). The first stage of construction was to begin in 2003, but the project became mired in controversy among various government agencies.[109] Critics point out that it already is possible to drive from Hualien to Taipei in about three hours, and traffic jams are common on the existing highway, especially on weekends and holidays; what value will be brought to the east coast, they argue, by encouraging even greater volume of motor traffic?[110] In spite of an effort by the Legislature (still under KMT dominance) to get the project back to the table for discussion at the end of 2005, the Executive Yuan (under the DPP) has postponed it indefinitely. A strategic environmental assessment of transportation needs in the northern and eastern regions of Taiwan was under

way in mid-2006, and was likely to show strong support for the highway from local residents. Regardless, current thinking among environmentalists is that this highway will end up not being constructed as long as the DPP government is in office, even though President Chen has come out in favour of the highway as long as it passes its EIA.[111] But that could change with the election in 2008, and the highway might yet be reinstated.[112] Environmentalists have pointed to the recently installed tilting trains as a viable alternative to the proposed freeway.[113] The Taroko Tilting Express (the name comes from the fact that the train literally tilts to counteract the centrifugal force when it rounds sharp curves) runs between Taipei and Hualien, shaving nearly an hour off the roughly three-hour trip, and there is discussion of expanding this system to lessen traffic congestion on roads.

3 *Large artificial lakes*
The Water Resources Agency is proposing to build four large artificial lakes (reservoirs) in Kaohsiung, Taoyuan, Tainan, and Yunlin Counties, to control flooding and to ensure water supplies in the local areas. Environmentalists have countered that nothing should be done on these until an island-wide comprehensive evaluation of water resources has been conducted, to prove the need for the lakes. While it is true that water shortages have become a problem in parts of Taiwan, especially the south-west coastal plain area, many environmentalists are yet unconvinced that artificial lakes are the solution. Moreover, much more careful study of the sites needs to be made, with complete EIAs carried out.[114] The Taoyuan Great Lake would cover 300 hectares. The GiYang Lake, which would involve damming the Kaoping River, would cover over 1,000 hectares but would provide only 340,000 tons of water daily for Kaohsiung and Pingtung residents. Critics argue that the 65 million cubic metres of dredged material would earn billions of dollars in profits for the contractors. Reallocation of water in the area, and less wasteful use of water, might well provide the same amount as this proposed lake, they also argue.[115] The 'Four Great Lakes' are probably a no-go as long as the DPP is in power. But again, changing political fortunes in 2008 might well tip the balance back to support for the lakes project.

4 *Large solid waste incinerators*
Large incinerators were being proposed, as part of the long-range plan to eventually do away with landfills. However, as noted earlier, the government's thinking about solid waste disposal is under review, given the public's growing opposition to both landfills and incinerators, and the reality that recycling has been so successful the existing incinerators around the island are not even being used to capacity. Hence, the likelihood of this project being carried out is small at this time.

5 *Hushan reservoir*
As part of the government's attempt to solve the island's growing water shortages, the Hushan Dam was proposed in Yunlin County. In this case, the project has become mired in controversy over a rare endangered bird

species that lives in the area, as well as some endangered plant species, and is likely to be a dead proposal now. The reservoir is part of the larger issue of the need for a comprehensive water resources study and more efficient use of the water already available, rather than building new reservoirs, according to environmentalists.[116] A similar situation revolved around the now-abandoned Meinung Dam project in Kaohsiung County. Not one of the ten new projects, Meinung was actually proposed in the 1990s, also to help solve the water needs in the south-west area in the 2001–21 period. Opposition to Meinung, however, forced the government to give up that project and propose improvements with existing reservoirs at Tsengwen and Akungtien. The four new proposed lakes (above) are the latest offspring of this effort. Solving the water problems of the south-west region remains one of the most difficult challenges facing the government, trying to placate environmentalists, aborigines, and other local residents, all worried about such things as ecology, ancestral rights, livelihoods, and related issues.[117]

6 *Cable-car projects in high mountains*
In 2004, the Council for Economic Planning and Development (CEPD), in an effort to boost both domestic and foreign tourism in Taiwan, proposed construction of major cable car systems on four of Taiwan's highest and most famous mountains: Yushan, Hsuehshan, Nanhushan, and Hohuanshan, in the three national parks of Taroko, Yushan, and Sheipa (see below). Citing the experience of other countries, such as Japan, and in Europe, which rely heavily on cable cars to get tourists up to higher elevations with minimal reliance on roads, the CEPD claims that cable cars are more environmentally friendly. At Yushan, Taiwan's highest peak, for example, according to government figures approximately 30,000 people reach the top of Jade Mountain each year (hardly an insignificant number for such a high peak) (see page 124). A cable car system would boost that figure to potentially 880 people per hour, or theoretically several hundred thousand people a year, and help boost annual visits to the three parks from the present five million to eight million. Environmentalists, led by various NGOs such as the Wilderness Society and student groups, were aghast at this plan and mounted public protests. Aside from the cost, critics charge that construction of the facilities would be destructive to the environment, and the huge increase in number of visitors would be very bad for the fragile high-country environments.[118]

Their protests seem to have worked. Those four projects appear to have been shelved. A fifth cable car system has also been proposed, to take visitors up Yangmingshan in the national park just north of Taipei (starting at the northern suburban hot springs town of Beitou and ending at the visitors centre in the park). There, supporters of the project feel, a highly utilized park bordering directly on Taiwan's major metropolitan region will have sufficient tourist revenues to warrant such a facility, and the risk to the environment theoretically might be much less than in more pristine, rural wilderness down south that would have been invaded by the other proposed cable car systems. Nonetheless, the Yangmingshan cable car project also is still bogged down in controversy. The opposition has been spearheaded by an NGO, the Wild At Heart Legal Defence Association, which claims that the Taipei City Government took over the project in 1998 and used its own Environmental Protection Bureau to exempt the project from an EIA, instead of turning to the TEPA. Then the BOT (build/operate/transfer) project proposal was turned over to the Yangmingshan National Park Administration, which was supposed to get approval from the Ministry of Interior. The critics were arguing that this administrative sleight of hand was a way to somehow circumvent the TEPA, which the government claimed was false.[119] Regardless, this project does demonstrate the reality that government bureaus and agencies at different levels sometimes have different agendas and do not always work harmoniously together.

7 *Reforestation projects*
Reforestation projects of various sorts were proposed, but as noted in Chapter 2 there is strong feeling in Taiwan now against clear cutting and elimination of natural forest.

8 *Reconstruction around the Central Cross-Island Highway*
One of the biggest controversies in Taiwan today revolves around what to do with the old Central Cross-Island Highway constructed in the 1960s but now partly out of use in major portions because of earthquake and typhoon damage. This important topic is covered separately in Chapter 4.

9 *Hydraulic development projects in Hualien County*
The government was proposing dams and connecting roads be constructed in Hualien County. However, almost all dam construction in Taiwan has been stopped, in light of the controversies noted earlier, with the exception of some smaller dams. In other words, Taiwan is going through some of the same soul-searching evident in the US now, regarding the value and wisdom of large dam and reservoir projects.

10 *Hydraulic projects in Tainan County*
Similar to the situation in Hualien County, except that the Nan-Hua Reservoir in Tainan has already been built. But future dam and reservoir construction is on hold for the foreseeable future.

In sum, the critics have been partially successful in stopping or delaying some of the so-called 'Ten Nightmares', but the controversy swirls on and will not soon end. There is likely always to be controversy over future major public works projects.

Nature conservation

Nature conservation is one area of environmental protection where Taiwan can justifiably take pride in what has been accomplished. In the space of just 30 years, Taiwan has managed to set aside about 20 per cent of the island's total land area in a multi-tiered conservation system consisting of 6 national parks, 19 nature reserves, 17 wildlife refuges, 31 major wildlife habitats, and six forest reserves. In addition, there are ten scenic areas, although technically these are not considered part of the 'true' conservation system, since these are developed for tourism only. This remarkable achievement has taken place in one of the world's most densely inhabited states, where intense land-use competition exists. For generations to come the people of Taiwan will benefit from this nature conservation system, which protects the natural splendour and flora and fauna of this unique island.[120]

The process actually started in the 1930s during the Japanese colonial era, when they established the first forest reserves and forest recreation areas. Then, during the martial-law era, the KMT government advanced the process by passing a series of acts: (1) the Statute for Development of the Tourism Industry in 1969, which led to the establishment of the first National Scenic Areas; (2) the National Parks Law in 1972, leading to the establishment of the first national park at Kenting in 1984; (3) the Cultural Heritage Preservation Law in 1982, which mandated the creation of a system of nature reserves and designated several species of rare and valuable plants for protection; and (4) the Wildlife Conservation Act of 1989 which further classified nearly 2,000 species of rare fauna into three levels of protection (endangered, rare and valuable, and requiring conservation measures). Those promising beginnings notwithstanding, the main nature conservation system has come into existence in the reform era since the late 1980s, with a series of laws and acts, in concert with the environmental movement in general, as detailed earlier in this volume (Figure 3.2).

National parks

The six national parks cover some nine per cent of Taiwan's land area, or more than 320,000 ha. Like many other countries in the world, Taiwan has modelled its national park system (and its nature conservation system as a whole) after that pioneered by the United States, with its establishment of Yellowstone National Park in 1872. The whole world conservation movement largely dates from then, and now there are about 1,700 national parks all over the world in 143 countries. Before World War II, the Japanese drafted plans for three parks at Alishan, Taroko, and Datun (Yangmingshan), but the war prevented putting those plans into action. Taiwan's national parks, like those elsewhere, serve as a microcosm of the island's topography, geology, climate, soils, river systems, scenery, animals and plants, cultural and historical landmarks. They are locations for both recreation and research. While they are designed first and foremost to preserve nature for future generations, they are not hands-off preserves forbidden to human use. Instead, that human use is carefully controlled and monitored to minimize change and damage to the environment. Initially, the government tried to use the BOT (build/operate/transfer) system[121] in the national parks, but that policy was

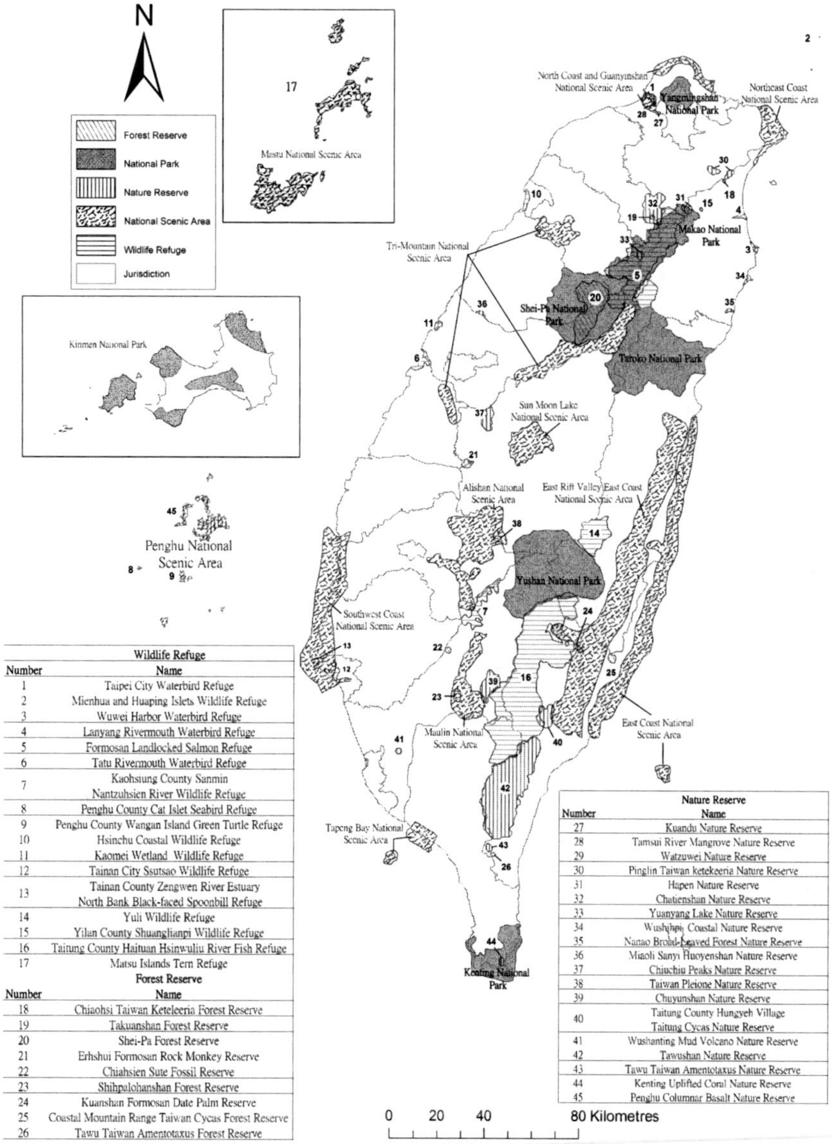

Number	Wildlife Refuge Name
1	Taipei City Waterbird Refuge
2	Mienhua and Huaping Islets Wildlife Refuge
3	Wuwei Harbor Waterbird Refuge
4	Lanyang Rivermouth Waterbird Refuge
5	Formosan Landlocked Salmon Refuge
6	Tatu Rivermouth Waterbird Refuge
7	Kaohsiung County Sanmin Nantzuhsien River Wildlife Refuge
8	Penghu County Cat Islet Seabird Refuge
9	Penghu County Wangan Island Green Turtle Refuge
10	Hsinchu Coastal Wildlife Refuge
11	Kaomei Wetland Wildlife Refuge
12	Tainan City Ssutsao Wildlife Refuge
13	Tainan County Zengwen River Estuary North Bank Black-faced Spoonbill Refuge
14	Yuli Wildlife Refuge
15	Yilan County Shuanglianpi Wildlife Refuge
16	Taitung County Haituan Hsinwuliu River Fish Refuge
17	Matsu Islands Tern Refuge

Number	Forest Reserve Name
18	Chiaohsi Taiwan Keteleeria Forest Reserve
19	Takuanshan Forest Reserve
20	Shei-Pa Forest Reserve
21	Erhshui Formosan Rock Monkey Reserve
22	Chiahsien Sute Fossil Reserve
23	Shihpalohanshan Forest Reserve
24	Kuanshan Formosan Date Palm Reserve
25	Coastal Mountain Range Taiwan Cycas Forest Reserve
26	Tawu Taiwan Amentotaxus Forest Reserve

Number	Nature Reserve Name
27	Kuandu Nature Reserve
28	Tamsui River Mangrove Nature Reserve
29	Watzuwei Nature Reserve
30	Pinglin Taiwan ketekeeria Nature Reserve
31	Hapen Nature Reserve
32	Chatienshan Nature Reserve
33	Yuanyang Lake Nature Reserve
34	Wushihpi Coastal Nature Reserve
35	Nanao Broad-Leaved Forest Nature Reserve
36	Miaoli Sanyi Huoyenshan Nature Reserve
37	Chiuchiu Peaks Nature Reserve
38	Taiwan Pleione Nature Reserve
39	Chuyunshan Nature Reserve
40	Taitung County Hungyeh Village Taitung Cycas Nature Reserve
41	Wushanting Mud Volcano Nature Reserve
42	Tawushan Nature Reserve
43	Tawu Taiwan Amentotaxus Nature Reserve
44	Kenting Uplifted Coral Nature Reserve
45	Penghu Columnar Basalt Nature Reserve

Figure 3.2 Nature conservation areas in Taiwan (source: Ministry of Interior, Dept. of National Parks).

stopped in 2004. Now the thinking is that facilities for visitors should all be outside the park boundaries (unlike the practice followed in the US), and the parks themselves should be strictly preserves. Environmental NGOs support this newer philosophy of management. A brief summary of each park's history and unique features follows.

Kenting

(1984) Covering over 33,000 ha, Taiwan's first park occupies the southernmost portion of the island, including the southernmost tip at Ouluanpi (Eluanbi) facing south toward the Philippines. This is a region of unique geography and tropical climate, with uplifted coral beds, ocean erosion, and fault line rifts, as well as unique flora and fauna. The park's longer existence, coastal location with easy road access, rich ocean life and coral reefs, plus extensive recreation and tourism facilities, make this one of the most frequented of Taiwan's six parks, especially popular in the winter when northern Taiwan can be rainy and cool.

Yushan

(1985) Taiwan's largest park, at over 105,000 ha, Yushan (Jade Mountain) is focused around the high core of Taiwan's central mountain range. The park contains 30 of Taiwan's tallest 100 peaks, including Yushan (3,952 metres), the highest peak in East Asia east of Tibet. Plate tectonics have produced some of the most rugged terrain in Taiwan, with steep ridges, cliffs, waterfalls, valleys and lakes. The park is home to the Bunun aboriginal tribe, and was traversed during the Qing Dynasty by the historic Batongguan Trail. Much of the park is accessible only by foot trail (something like the high mountain country in Yosemite); hence, this remains one of the less visited parks. Nonetheless, some 300,000 people annually file applications for entry to Yushan, although a much smaller number actually enter the park.[122] As noted in Chapter 2, this area was saved by environmentalists from having a cross-island highway slashed through it. They also succeeded in stopping plans for a cable car tourism development of Yushan (see above). Environmentalists want minimal tourist facilities in this park, perhaps in part because the mountain is playing an increasing role in the search for icons of Taiwan's new nationalist spirit, by both native Taiwanese as well as the indigenous Bunun aborigines.[123]

Yangmingshan

(1985) Taiwan's second smallest national park, at about 11,500 ha, Yangmingshan is the one park sited closest to a major urban area (Taipei) and the only one located along a system of dormant volcanoes. The park's craters, sulphur vents, and hot springs have long been of interest to people, including the Japanese, who started development of the area. The park's unique geologic history and proximity to more than six million people living in the Taipei metro region make this the most visited of Taiwan's parks. In 2006, more than 1 million people visited the park just during the first three weeks of the flower-viewing season, which lasts about one month from late February to late March. Up to 120,000 people clogged the roads each day on weekends.[124] Diverse vegetation, including beautiful flowering plants, numerous trails and tourist facilities, and good road access, draw these millions of visitors each year. Somewhat like

Central Park in New York City, Yangmingshan serves as a primary recreation site for residents of the Taipei area.

Taroko

(1986) Located on the east coast in Hualien County, with 92,000 ha, and considered by many to be Taiwan's most spectacular natural site, Taroko National Park is built around a great natural gorge that the Liwu River cuts through a massive marble formation. Added to that is a spectacular road, begun by the Japanese and finished by the KMT government in the 1960s as part of the Central Cross-Island Highway (see Chapters 2 and 4), that winds through the marble canyon. The park also contains 27 peaks rising over 3,000 metres to create a stunning mountain landscape that rightly ranks as one of Asia's great natural wonders and a top tourist attraction in Taiwan. Many foreign visitors come to Taiwan just to see Taroko and then leave.

The park also is home to the Atayal aboriginal tribe, who have been asserting their 'rights' as the government attempts to develop further tourist facilities in the park. The controversy as of 2006 was over efforts to build a luxury hot spring hotel in the park, without first consulting the tribe. The Taroko Tribal Autonomy Promotion Committee is opposed to the idea, and wants the government to allow the Taroko Tribe to co-manage the park.[125]

Shei-pa (i.e. Hsueh-pa)[126]

(1992) Situated in the high mountains of north central Taiwan, Shei-pa, at just under 77,000 ha, is a high alpine park something like Yushan. Bounded on the south side by the famous upper Tachia River valley and the Central Cross-Island Highway (see Chapter 4), most visitors see only this side from the road. Much of the park is accessible only by foot trail, and hence the interior sees relatively few visitors. Besides spectacular mountain vistas, the park preserves a rich variety of flora and fauna in climate zones ranging from tundra to warm temperate. The Atayal and Saisiyat aboriginal groups have lived here for centuries. This park, along with Yushan and Taroko, best preserves the mountain grandeur of interior Taiwan.

Kinmen (i.e. Chin-men)

(1995) Kinmen is the most unique of the national parks, in that it is located on Kinmen Island, just 8 km offshore of Fujian Province in China, some 277 km west of the island of Taiwan. Kinmen (consisting of two main islands) is one of the island groups held onto by the retreating ROC forces in 1949 when the Communists took over the mainland, and for decades was nothing but a massively fortified, restricted outpost in the ongoing struggle between Taiwan and the PRC. With the lessening of tensions between the two sides in the 1990s, the government opted to turn parts of the two islands into a national park, as a way of building up the local economy, while preserving the island's unusual natural features and unique history. With less than 4,000 ha, the national park occupies one quarter of Kinmen. Historical preservation is a prime objective here, and visitors can see many of the famous sites where battles were fought between ROC and PRC forces in the past, and where ROC forces were entrenched to hold off future invasion that never came, as well as restored villages and structures dating back well into the Qing Dynasty, sights impossible to find in Taiwan itself now. Today, hordes of visitors from Taiwan (and other parts of Asia) as well as China come to Kinmen to sightsee, shop, and eat, in a rather bizarre setting.

Makao (Magao)

A seventh national park may be in the works, but is not yet formalized. Makao has been proposed for over 53,000 ha stretching across the mountain land connecting Ilan, Taipei, Taoyuan, and Hsinchu Counties, just north of Sheipa and Taroko National Parks. Environmental activists and members of the Atayal tribe who live in that area first proposed the idea for this park in the 1990s, with the primary goal of preserving the 900 ha of endangered and valuable cypress forest found there. Opponents, however, who include some aborigines, point out that the park area is a traditional aboriginal hunting area and hunting would be restricted by the formation of a national park. The government is trying to negotiate with the aborigines to work out a co-management system that would preserve aboriginal culture while at the same time providing new economic opportunities, through tourism, in an economically depressed area.[127] The issue is tangled up with a draft law that would allow the formation of indigenous people's autonomous areas, and there is controversy as to which law would have precedence – the autonomous areas law or the national park law. The outcome of this could have far-reaching consequences for the whole nature conservation system in the mountain interior (see Chapter 4 also). The thinking among environmentalists as of late-2006 was that Makao was not likely to be approved, because of ongoing opposition of aboriginal legislators, further evidence of the growing assertiveness and power of the re-awakened aborigines in Taiwan.

Dongsha Marine National Park

The seventh national park actually may turn out to not even be on Taiwan Island itself. In late 2006 the government's Construction and Planning Administration (responsible for all National Parks) announced formal plans to create the Dongsha Marine National Park around Dongsha Island (aka Pratas Island) in the South China Sea, some 240 nautical miles south-west of Kaohsiung. This would be Taiwan's first marine park, and its largest by far, covering some 80,000 ha of atolls and around 350,000 ha in administrative area (mostly sea). The park ostensibly is designed to preserve the area's unique marine ecology with its coral atolls.[128] The unspoken objective, clearly, is to stake a firmer claim to Dongsha, which is an area also claimed by the PRC, part of the larger South China Sea region which is contended by the various riparian nations (China, Taiwan, the Philippines, Malaysia, Indonesia, and Vietnam), focused especially around the Spratly Islands further south. Hence, this park is likely to be fraught with much controversy, if it actually is established. Not surprisingly, the government has already warned that tourists will not be allowed to visit the park for at least ten years because there are no plans to establish tourist facilities soon. Rather, the focus would be to develop a scientific research station and make the island 'self-sufficient', meaning inhabitable by Taiwan military and scientific personnel, which theoretically would strengthen Taiwan's claim to Dongsha under the 1982 Law of the Sea agreement.

Nature reserves

Taiwan now has 19 nature reserves covering nearly 65,000 ha, ranging in size from a mere 5 ha plot to protect volcanic land forms in Kaohsiung to the 47,000 ha forest reserve surrounding Tawu Mountain in Pingtung County. Twelve of these reserves are directly managed by the Taiwan Forestry Bureau under the Council of Agriculture (COA); the remaining seven are managed by local government agencies and the Taiwan Forestry Research Institute, which all report to the COA. All of these reserves were established to maintain the proper functioning of ecosystems and biodiversity and to ensure sustainable use of biological resources. Thus, they protect rare or endangered species or ecosystems, while providing reasonable recreation areas for people also.

Two of the oldest reserves are in the Taipei Metro Region, and played roles in the early stages of the environmental movement of the 1970s and 1980s. The Tamsui River Mangrove Nature Reserve, established in 1976, with just 76 ha, is situated near the mouth of the Tamshui River, and preserves a typical estuary ecosystem of mangroves and other plants in which large numbers of indigenous and migratory birds nest. The Kuandu Nature Reserve was set up ten years later, in 1986, on 55 ha of wetland at the confluence of the Keelung and Tanshui Rivers. Reeds and mangrove dominate, and the area is rich with migratory birds and a great variety of marine life. Both of these reserves are popular with residents of the Taipei area.

Wildlife refuges and habitats

Wildlife refuges and major wildlife habitats are set up to protect wild animals such as sea turtles, migratory birds, waterfowl, freshwater fish, and their habitats. Since 1991, with approval from the Council of Agriculture, local governments have established 17 wildlife refuges, and the COA has designated 31 sites as major wildlife habitats. The wildlife refuges cover a total of over 25,000 ha, while the habitats cover a much larger total area of just under 300,000 ha. Both are located in coastal as well as mountain interior sites all over the island. As noted in Chapter 2, Taiwan's fauna are under intense pressure from human occupance, and these reserves and habitats are vital to preventing the extinction of endangered species. The test is to prove that it is possible for both humans and fauna to co-exist on this crowded island when proper safeguards are put in place.

National scenic areas

National scenic areas are, in a way, the first stage in the process of setting up a national park. There are currently ten scenic areas set aside, covering almost 470,000 ha, the largest single category of nature conservation areas in Taiwan. Scenic areas are designed to strike a balance between protection of natural landscape and tourism promotion. Kenting National Park, for example, began as a scenic area in 1979, before becoming a formal national park in 1984. Some of the most famous recreation and tourism sites in Taiwan are now designated scenic areas, including Mount Ali (Ali Shan), Sun Moon Lake, Taiwan's northeast coast (where some of the best bathing beaches are located), the Penghu Islands, and the rugged east coast, among others. It is possible that one or more of these scenic areas might be elevated to national park status in coming years. This is a complex process, though, because national park status puts additional restrictions on tourism and tourist facilities, to minimize the negative impact of visitors.

A prime example of a region that absolutely needs protection while it is still in a relatively unspoiled condition is the east coast, in Hualien and Taitung Counties. Two national scenic areas have been created there. The first one, set up in 1987, covers over 41,000 ha of the east coast, home of the Ami tribe and a largely wild and undeveloped 176 km-long strip with steep mountains rising out of the sea, magnificent seascapes, rugged beaches, impressive geologic features, and unique flora and fauna. White water rafting is popular on the Hsiukuluan River that flows here. The scenic area also includes off-shore Green Island, once notorious as a political prison site during the martial law era.[129] The other scenic area was set up in 1997 to save the beautiful Eastern Rift Valley, a 158 km narrow corridor of more than 138,000 ha between the Central Range and the East Coastal Range. Visitors can see picturesque farms, waterfalls, and hot springs, among other things.[130] This was a frontier region not long ago, opened up by the Japanese during the colonial era and among the last to be settled by Taiwanese farmers. The Japanese built the first railway here, a narrow-gauge line running between Hualien and Taitung, the two cities that anchor either end of the valley. Several aboriginal groups are found here, and the population density is still very light compared to the west coast of Taiwan. The total population in both the valley and coast is only a few hundred thousand. The region, in some ways, resembles the Taiwan of the early twentieth century, when life was simpler, less crowded, and the environment still clean. Air and water pollution, for example, are relatively minor problems. The region is now easily accessible by various transport modes, including air, so tourism has great potential but must be carefully managed to protect the environment.

In late 2006 the government announced the launch of a 'mega-development project' in eastern Taiwan, the so-called 'Eastern Taiwan Sustainable Development Plan'.[131] All the key agencies are involved in the project: Ministries of Transportation and Communications, Interior, Economic Affairs, the Council of Agriculture, and the TEPA. The plan is to spend US$1.5 billion over a period of nine years to transform the eastern corridor into an 'international-standard' tourism area, while also addressing other economic needs. Specific parts of the plan include: developing tourism in the region's rural townships; establishing an industrial processing zone for deep-sea products; promoting an 'immigration' programme (meaning encouraging in-migration from western Taiwan) that would include building new villages; developing a cottage industry-based economy; strengthening the railway and other transport systems; and reviewing (stopping?) the existing policy of shipping sand and gravel from eastern Taiwan to the west for construction needs. This project obviously is designed to give eastern Taiwan a major boost in development efforts, but to do so in a sustainable way that will theoretically protect the environment. It is an ambitious undertaking, fraught with great dangers as well as opportunities. Those who favour keeping eastern Taiwan as pristine and undeveloped as possible are likely to be disappointed with this endeavour.

Forest reserves

National forests account for 58 per cent of Taiwan's total land area today. Of that, about 72 per cent of the 1.6 million ha of national forestland are 'natural' forests, i.e. relatively undisturbed. Forest reserves thus are national forest lands that possess unique characteristics the government wants to preserve, and hence preservation is emphasized over development. Typically, forest reserves later become elevated to nature reserve status. There are now six forest reserves, totalling under 22,000 ha, maintained by the Forestry Bureau under the Council of Agriculture. The government does allow limited 'educational' tourism in these reserves, with a network of hiking trails and hostels, open to the public for a fee, and offering fairly rustic accommodations to those wanting to experience that kind of environment.

In sum, there is in place now a fairly rich hierarchy of nature conservation areas in Taiwan that provide varying levels of recreation and opportunities to enjoy nature, ranging from comfortable, well-developed tourist facilities to fairly primitive and remote facilities. The beneficiaries are all the people of Taiwan, whose quality of life is significantly enhanced by this system, even if one never actually visits these sites. It is enough for some just to know that such preservation is taking place. Most people on the island do, however, visit these various conservation areas at some time, as millions of site visits take place year-round. The main danger now is from over-use, in the syndrome of Yellowstone/Yosemite/Grand Canyon (in the US), and the need at some point to set limits on numbers of visitors per day and other restrictions in some of the conservation areas.

Energy and Taiwan's role in curbing global emissions

At present, Taiwan's energy comes mostly from four sources – coal, natural gas, oil, and electricity. The latter two account for about 90 per cent of total energy consumption. In the 20 years between 1984 and 2004, oil dropped from 50 to 40 per cent of consumption, while electricity increased from 30 to 50 per cent. Industry still consumes the largest share, 57 per cent, but that is a slight decrease from the 60 per cent in 1984. Agriculture is much less important now to the economy and consumes less energy. Total energy consumption has risen an average of 6 per cent a year.[132] Another way to look at this is in terms of total energy supply, which in 2005 consisted of 51.3 per cent from petroleum, 31.9 per cent from coal, and the remaining modest amounts from natural gas (8 per cent), nuclear power (7.3 per cent) and hydro power (1.4 per cent).[133] Per capita electricity consumption surged from 2,595 kwh in 1984 to 9,131 kwh in 2004, a 3.5-fold increase.[134] Total energy consumption amounted to over 53 million KLOE in the first half of 2006, a 2.5 per cent increase over 2005.[135] Like just about every nation or state in the world today, Taiwan faces the dilemma of trying to produce enough energy to meet its voracious appetite, and do so in an environmentally friendly or stable way. Moreover, Taiwan must accomplish this while simultaneously doing its part to curb global emissions. Some might argue the two goals are inherently contradictory and impossible. Part of the dilemma is in getting a consensus on what exactly constitutes an environmentally 'friendly' source of energy. Is it thermal plants fired by coal, oil, or natural gas? Is it hydropower? Is it nuclear power? Is it so-called 'exotic' or renewable energy (wind, geothermal, tidal, bio-energy, etc.)? Taiwan has not yet found that consensus, and given the political atmosphere on the island is not likely to soon achieve a consensus (see also Chapters 2 and 5).

One projection for energy sources from the Bureau of Energy is shown in Table 3.3. In their scenario, coal use will increase significantly, basically trading places with oil, between 2010 and 2025. Natural gas use will also increase, but hydropower will fall sharply in its share. Nuclear power will slip slightly, but already be much lower by 2010, while renewable energy will double in those 15 years.

Table 3.3 Projected primary energy demand, 2010 and 2025

Energy source	*2010 (%)*	*2025 (%)*
Coal	30.7	42.4
Crude oil	42.9	29.5
Natural gas	12.7	17.4
Hydropower	9.6	3.6
Nuclear power	0.7	0.5
Renewable energy	3.3	6.6

Source: Bureau of Energy, Ministry of Economic Affairs (as reported in *Taiwan Review*, April 2006, p. 8).

However, other projections, with different figures, have been issued by the government from time to time, so the future mix of energy sources in Taiwan remains fluid. For example, in 2005 the government declared its intention to stick with its pledge to make Taiwan nuclear-free within 40 years, as the nuclear plants are phased out. The hope was to have 10 per cent of the nation's power generated by renewable-energy sources (especially wind power, but also solar energy and biomass) by 2010 and 12 per cent by 2020.[136] Nuclear power accounted for under 18 per cent of the island's installed electricity generating capacity as of early 2005, down significantly from its share back in the mid-1980s. This was due to greatly increased production of power from coal- and oil-fired plants. The plan was to increase the capacity of coal-fired plants by some 65 per cent by 2014, which would make such plants account for 36 per cent of Taiwan's power capacity then, the two main sources of coal being Indonesia and Australia.[137]

And yet, even thermal electric plants create controversy. Taipower has proposed building a new coal-fired thermal plant, with two generators, in Changhua County. Critics have complained, however, about the projected yearly emission of 8 million tons of CO_2, which would be a violation of the Kyoto protocol, and also about higher rates of lung and liver cancer reported in that coastal area from Taichung to Tainan where the plant would be located.[138] The government seems to be caught in another no-win situation on the energy front. To reduce reliance on both thermal and nuclear energy, as noted above, the government stated its intention that Taiwan should generate 10 per cent of its electricity from renewable sources by 2010, almost a doubling from the current figure of 5.5 per cent. Wind power is increasingly favoured now. Taipower currently has about 100 wind turbines in operation along the west coast and is planning to build another 200 on land in Taiwan and the Penghu Islands before 2010. After that, another 546 turbines will be built offshore in shallow coastal waters along the west coast, in an ambitious scheme to boost wind power generation further between 2010–20.[139] The Ministry of Economic Affairs even predicts that eventually (target date unspecified) wind energy will account for 35 per cent of all renewable energy in Taiwan.[140] Clearly, different branches of government are issuing different projections, indicative of a lack of solid agreement on how to proceed.

At the same time, the government is trying to convince energy users to conserve energy as much as possible. For example, air conditioning in Taiwan's long, torrid summer consumes vast amounts of electricity. So everyone is being urged to turn up the thermostat and dress lighter, including office workers who are encouraged to do without a suit and tie as much as possible.[141] This is easier said than done in Taiwan's more formal culture than in other more casual cultures.

As for the issue of disposing of nuclear waste, the plan, as noted in Chapter 2, had been for the government to spend five years in choosing a new site away from the existing one on Lanyu Island, and another four years for the 97,000 barrels stored on Lanyu to be inspected, repacked and removed, and then four years more for the empty storage site to be rehabilitated. In addition, as of mid-2006 there were 35,000 barrels of low-level radioactive waste stored at the First Nuclear Power Plant, and more than 44,000 barrels at the Second Plant, which also uses supercompactors and incinerators to compact or burn waste.[142] The government now thinks that the Lanyu waste site can be decommissioned by 2014. However, this scenario assumes that a suitable alternative site, inside or out of Taiwan, can be found soon.[143] Taiwan officials have even visited the Yucca Mountain Repository in Nevada, to see how the US is handling the problem of nuclear waste disposal.[144] Unfortunately, Taiwan does not have the abundant space that the US has.

Regardless, the current worldwide trend is back toward favourable views on nuclear power in most countries, caused in part by ever rising oil prices and the instability of oil markets, with one report predicting that 800 new nuclear reactors will be built worldwide in the next 25–35 years and might even reach 1,200–1,500 new reactors.[145] Another report predicts doubling of world energy demand by 2050, with 1,000 nuclear power plants in operation around the world, compared with fewer than 450 today.[146] Given that trend, it is possible to visualize Taiwan swinging back to rejoin the nuclear energy bandwagon at some point down the road. It may be that Taiwan will never even give up nuclear energy. That may be even more likely if somehow Taiwan is reunified with the PRC in this century, as China has wholeheartedly embraced nuclear power. However, as of mid-2006 public opinion in Taiwan, especially among NGOs, and irrespective of Pan-Blue or Pan-Green parties, was strongly against nuclear power.[147] Hence, Taiwan appears to be bucking the current world trend for the time being at least.

On the other hand, the desire and need to more effectively curb atmospheric emissions (i.e. greenhouse gases, or GHG) might be one factor that could help push Taiwan back toward a pro-nuclear energy policy in coming years. As of 2005, Taiwan ranked 22nd in the world in emission of carbon dioxide, a mere 1 per cent of annual industrial emissions worldwide. Nonetheless, Taiwan's proportion keeps inching upward. Carbon dioxide emissions totalled 270 million tons in 2004, accounting for 74 per cent of the island's GHG. That figure is predicted to grow to 530 million tons over the next two decades if Taiwan does nothing, an annual growth rate of about 6 per cent. The TEPA has even said that the figure could go as high as 665 million tons by 2020.[148] Most of the CO_2 comes from motor vehicles and coal-fired thermal plants. LNG produces about one-third to one-half less GHG per unit of electricity than a coal-fired plant, so Taiwan is looking into ways to switch to use of LNG in thermal plants. Although not a signatory to the Kyoto protocol, Taiwan plans to cut carbon dioxide emissions by a total of 170 million tons by 2025.[149]

Achieving that goal will not be easy nor inexpensive. When the Kyoto Protocol became effective in February 2005, it addressed the emission of six kinds of GHG that contribute to global warming, most importantly CO_2, and imposed legally binding caps on GHG emissions in 35 developed nations, mainly from burning fossil fuels in power plants, factories, and cars. Unfortunately, the US, India, China, and Brazil, among others, were not signatories. The participating developed nations must cut emissions of GHG by 5.2 per cent below 1990 levels by 2012.[150] Taiwan, along with other developing nations, is part of the 'second wave' of the protocol's implementation process, even though Taiwan is not a formal signatory to the pact. In 1998, after the Kyoto Protocol was first signed (1997), Taiwan held a National Energy Conference, which led to passage of the Statute on the Development of Renewable Energy, with the aim of ensuring that at least 3 per cent of Taiwan's electrical power would be generated by use of low-pollution renewable energy by 2020.

More recently, concerned that Taiwan might end up becoming a target of criticism or even trade sanctions, and ever mindful of the need to maintain a good international image, with China breathing down its neck, Taiwan is trying to cooperate even more fully with the directives of the Kyoto Protocol, and held a second National Energy Conference in June 2005, pledging to accelerate the pace for boosting the amount of the nation's electricity needs coming from renewable energy resources by 2010.[151] Taiwan's leaders also have urged the U.N. to include Taiwan in future talks in the Protocol process, but the PRC is likely to resist such a move (for obvious political reasons), just as it has consistently blocked Taiwan's participation in the World Health Organization.[152] Regardless, Taiwan initially set 2012–20, as the period when it would do its part for reducing global emissions. The TEPA announced that Taiwan would target a reduction of carbon dioxide emissions by about 36 million tons a year by 2020.[153] While that would be a step in the right direction, GHG emissions then would still be much higher than they are now. One of the NGOs, the Taiwan Climate Protection Coalition, proposed that Taiwan be much more aggressive about curbing GHG, cutting emissions to the 2005 level by 2015. One way to do this would be for Taiwan to participate in the Kyoto Protocol's 'Clean Development Mechanism' (CDM), which allows developed nations to buy GHG emission quotas from developing nations, in effect transferring wealth from richer to poorer nations. Taiwan thus theoretically could invest in CDM projects in the PRC, for example.[154]

There is plenty of evidence, though, that some forces in Taiwan could work against the long-range goals of reducing GHG emissions.[155] A notable recent example was the proposal by Formosa Plastics Group to build a steel-making plant, designed to produce ten million tons of steel annually, but which would emit an estimated 20 million tons of CO_2 per year.[156] This steel project was still in EIA process as of 2006 (see Chapter 5). What is the government to do? It needs economic growth, yet needs to curb emissions.

In September 2006, the Executive Yuan unveiled a bill that proposes making serious cuts in emissions of carbon dioxide and five other gases. Called the 'Law for the Reduction of Greenhouse Gas Emissions', if passed into law it would fall to the TEPA to draw up specific reduction strategies, including quotas, monitoring, penalties, etc.[157] The TEPA would assume enormous authority in the arena of GHG. The bill must first be passed in the Legislative Yuan, a process that could take quite some time given the current state of politics on the island (in fact, passage before the 2008 presidential election would seem highly problematic). One could anticipate major industries lobbying quite strongly against the bill.

Conclusion

Perhaps an appropriate way to conclude this chapter is to note the results of a public survey of 1,097 Taiwanese conducted by the Environmental Quality Protection Foundation, an NGO, in January, 2006. The responses were revealing:[158]

1 Public confidence in environmental protection in Taiwan decreased two points on a scale of 100 in 2005.
2 Some 37 per cent thought that recent efforts to improve the local ecosystem were disappointing, and 63 per cent of those who felt Taiwan's environmental quality had worsened were better educated. Just 26 per cent supported the then current TEPA chief, Chang Kuo-lung.
3 Curiously, the most worrisome environmental issue was that of cellular phone antenna sites installed on apartment roofs, the interviewees apparently concerned about the impact on health from the radiation. A majority, 64 per cent, thought the government should require such installations to be located only in non-residential areas.
4 Over half believed that the TEPA's three-year policy of restricting plastic-made products had been a 'total failure'.[159]
5 Another high visibility issue was the dioxin residues found in local dairy foods, as well as in duck meat and eggs from Changhua County, which surfaced in the summer of 2005. Nearly 90 per cent thought the government should draw up stricter standards for dioxin content in all foods.

These survey results show again that the public responds to environmental issues not entirely on the basis of factual knowledge but also from perception, personal experience, media reports, and imagination. Hence, the government's task, through the TEPA and other organizations, is to provide accurate information and to educate the public. It is vitally important that transparency be a guiding principle of environmental protection, because in the absence of hard facts imagination can take over. Successful environmental protection requires the direct participation and support of the public, more so than almost any other government initiative. The TEPA (and government in general) has not yet won this battle.[160]

4 Whose land is it?

Land use issues

Land is precious in Taiwan. Thus, competition for use of land can be intense and lead to serious environmental problems, as was introduced in the preceding chapters. This intense land use competition raises challenging questions about who exactly owns the land in Taiwan and who has the right to dictate how land should be used. Is it the government? The aborigines? Private individuals or businesses? Society as a whole? The issues facing land use are the primary focus of this chapter, which is divided into two main parts. The first part looks at land use in the mountain areas and slopelands that account for nearly three-quarters of Taiwan, and does so through a detailed examination of the seminal project that opened up the interior to settlement and development, the first cross-island highway, and its impact. The second part looks at land use in the lowlands of Taiwan, the main agricultural area, and the critical issues facing that region, especially from an environmental viewpoint, since Taiwan joined the WTO.

Land use issues in Taiwan's mountains: the Central Cross-Island Highway and its impact[1]

Although efforts to penetrate and develop Taiwan's mountain interior have been going on since the earliest history of the island, one could say that the seminal project of the ROC (post-Japanese) era after World War II that really gave a green light to opening up the interior was the first cross-island highway. Some 50 years later, there now is widespread agreement in Taiwan that the highway is a prime example of unwise use of mountain slopelands and its consequences. This was not always the case; when it was first planned and constructed the road was widely touted as a great engineering feat. Now, this famous public works project is seen as a notorious example of the political environment during the martial law era (pre-1987) when the government could do just about whatever it liked, without worrying about environmental impact assessments, public hearings, or any other checks and balances (see Chapter 1), in marked contrast to the experience of the later cross-island highways and other more recent public works projects, which have had to face the scrutiny of environmentalists and activists since the 1980s and on.[2]

The upper Tachia River basin

The focal point of the first cross-island highway became the upper Tachia River basin in central Taiwan. The Tachia is one of Taiwan's major rivers, flowing out of the north central mountains with its headwaters in easternmost Taichung County and contiguous areas. The fourth longest river in Taiwan, the Tachia enters the Taiwan Strait off the Taichung coast some 123 km from its source[3] (see Figure 1.3). When the Japanese arrived in Taiwan in 1895, the upper Tachia basin was very lightly inhabited, remote and inaccessible from the lowland. The Atayal aborigines in the area practised hunting and gathering and limited shifting cultivation. Japanese policy in their 50 years of rule fluctuated between conservation and efforts at exploitation of the mountain resources, and between assimilation and suppression of the aborigines. The mountain interior, more than one-half of Taiwan, was set aside as a Special Aboriginal District, with strictly controlled entrance and exit, but the Japanese also permitted licensed logging operations in the area as they brought it under their control. The sheer difficulty, however, of reaching the upper Tachia basin more or less protected the valley from serious exploitation. The Japanese hacked a foot track through the rugged gorge in the middle reaches of the river, but it was inadequate for motorized access and any real kind of development effort. Hydrologic surveys of the river were carried out, and plans for dams, including one at the entrance to the upper basin, were drawn up, but none were built. This was partly due to the fact that the Atayal were among the least cooperative aborigines for assimilation, and hence real Japanese control of the upper basin was not achieved until the 1930s, and by then Japan was preoccupied with its militarization and expansion into China and Asia. Had the war turned out differently, and Japan stayed in Taiwan, it is possible that development of the high mountain country might have proceeded with vigour in later decades. However, the Japanese abandoned the idea of a cross-island highway in the pre-war years because they thought it infeasible.[4]

With the retreat of the ROC government to Taiwan in 1949, and subsequent efforts at economic development of the island, attention was once again focused on the interior. The tremendous population increase of the island (which reached over eight million by the end of 1952), massive industrialization plans with attendant need for large increases in electricity generation, the need for lumber, the desire to link the two sides of the island, and the need to find productive work for thousands of ROC soldiers, all these factors and others converged on the upper Tachia basin, as the government made the fateful decision to build a highway through this area. Without adequate transportation access, the region's resources and development potential would remain locked up.

The basin's resources offered the potential for a variety of uses. The river itself is one of Taiwan's largest in terms of annual discharge. Just below the upper basin, the river flows through a rugged, narrow gorge before entering the west coast plain where it broadens into a shallow, braided stream typical of Taiwan's rivers in their lower reaches. The large annual discharge and steep gradient made the Tachia River, sometimes called 'Taiwan's TVA', a natural site for hydroelectric power production. The forest cover had long been a timber resource. Even though slopes are steep and soils shallow, the gentle climate made the basin attractive as a site for growing temperate fruits and vegetables that simply could not survive in the torrid lowlands. Moreover, the spectacular mountain setting gave the area strong tourism appeal. All these assets combined convinced the government to build the highway and open this area to settlement and development.

The Tachia River valley was the natural route for the new highway to take, as the road was laid out across the high and extremely rugged mountains of central Taiwan. The upper basin of the river covers some 1,270 km^2 (over 61,000 ha) in the upper slopeland zone of the island, with local relief ranging from 1,230 metres at the entrance to the upper basin, to a maximum elevation of 3,884 metres at the highest peak (Hsueh Shan or 'Snowy Mountain'). The landscape is extremely rugged, with steep slopes prevalent. The climate at this elevation is temperate, with high annual precipitation of 2,500–3,000 mm, concentrated in the summer months. Originally, the basin was entirely covered by dense coniferous and mixed hardwood/coniferous forest typical of the high mountain country of Taiwan.

The cross-island highway was begun in 1956 by personnel in the Vocational Assistance Commission for Retired Servicemen (VACRS), under the leadership then of Chiang Ching-kuo (Chiang Kai-shek's son, who went on later to succeed his father as President of the ROC), and completed in 1960, at a cost of 212 workers killed, and 780 injured.[5] This first cross-island highway was seen not only as a practical way to provide employment to retired servicemen (mostly Mainlanders), but symbolically as a demonstration of the ability of the government and people of Taiwan to conquer nature. At that time in Taiwan's history, the island was still relatively poor and needed to exploit local natural resources to build its economy. Specifically, thus, the road would meet a number of critical needs:

1 The road would link the populous west coast plain with the lightly populated, greatly under-developed eastern side of the island, then accessible by land only via a narrow, dangerous road down the north-east coast carved into the cliffs between Suao and Hualien during the Japanese era.
2 The road would make it possible to build and service hydroelectric dams at the entrance to the upper valley as well as downstream, generating much-needed electricity and helping control flooding of the Tachia River.
3 The road would provide tourists (domestic and foreign) access to parts of the Central Mountains previously almost unreachable, and open up one of the most scenically splendid mountain environments to be found anywhere in Asia, one largely unknown to the outside world.
4 The road would provide access to mountain slopelands in a more temperate climate regime to give VACRS the opportunity to experiment with high-altitude temperate agricultural production of fruits and vegetables.

5 The road's construction would provide jobs for military personnel and then many of them also could be settled on large state farms in the area to undertake agricultural cultivation, thus partially solving an ongoing problem for the government – what to do with retired mainland soldiers, most of whom fled to Taiwan without wives or families?

6 The road would provide enhanced economic opportunities for the resident Atayal aboriginal people to grow crops for sale in lowland markets, to get involved in commercial tourism ventures, and to have enhanced access to the modern society developing in the plains.

In short, many people at that time believed the road was a great idea and would help advance Taiwan. Few, if any, foresaw the unintended consequences.

The project was bold and imaginative, but dangerous, resulting in a twisting two-lane road (to call it a 'highway' was a bit of an exaggeration) that partially followed the Hohuan Old Trail (aka Nengkao Old Trail), a major east–west foot trail developed over the previous century during the Qing Dynasty as a route through the formidable mountain interior.[6] The highway proceeded from near sea level to a maximum elevation of 3,200 m. at its highest point along the 192 km length of the highway (Figure 4.1). A 112 km spur was also built leading off from Lishan,[7] the main town in the upper basin, to the north-east coastal port of Ilan. In addition, a 42 km spur (initially, a difficult tract suitable only for four-wheel drive vehicles) was constructed south-west from Tayuling to the town of Wushe, with connections there to roads going on to the west coast plain. Lishan and the upper basin thus became the pivotal node for cross-island traffic running in four directions. Two other cross-island highways were later constructed (the Northern Highway, from Taoyuan to Ilan, and the Southern Highway, from Yuching in Tainan County to Haituan in Taitung County), but neither of those roads had the extraordinary impact that the first one had. Rising concerns by environmentalists by the 1980s convinced the government to postpone (and eventually cancel) any plans for further cross-island highways.

After the road opened for business, under the best of circumstances the trip across in either direction took the better part of a day to make by bus or car, and was definitely a scenic wonder and thrill ride, but not for the faint of heart. Building the road required making steep cuts on mountain sides, plus the digging of some 38 tunnels along the route, with the excavated material just being dumped in the valley below with no concern about its impact on drainage, flora, or fauna. Slopes alongside the highway in many places were 45–60°, considerably greater than what is now considered a safe angle of repose.[8]

The result was that over the years landslides became a common occurrence along the highway, with the road often closed for periods of time while necessary repairs were made. The government diligently invested large amounts of money and effort to maintain the road down through the years, because of the benefits perceived to be derived. In many ways, the upper Tachia River valley and the cross-island highway became the symbol of the whole effort of trying to develop the resources of the mountain interior of Taiwan. It was here where

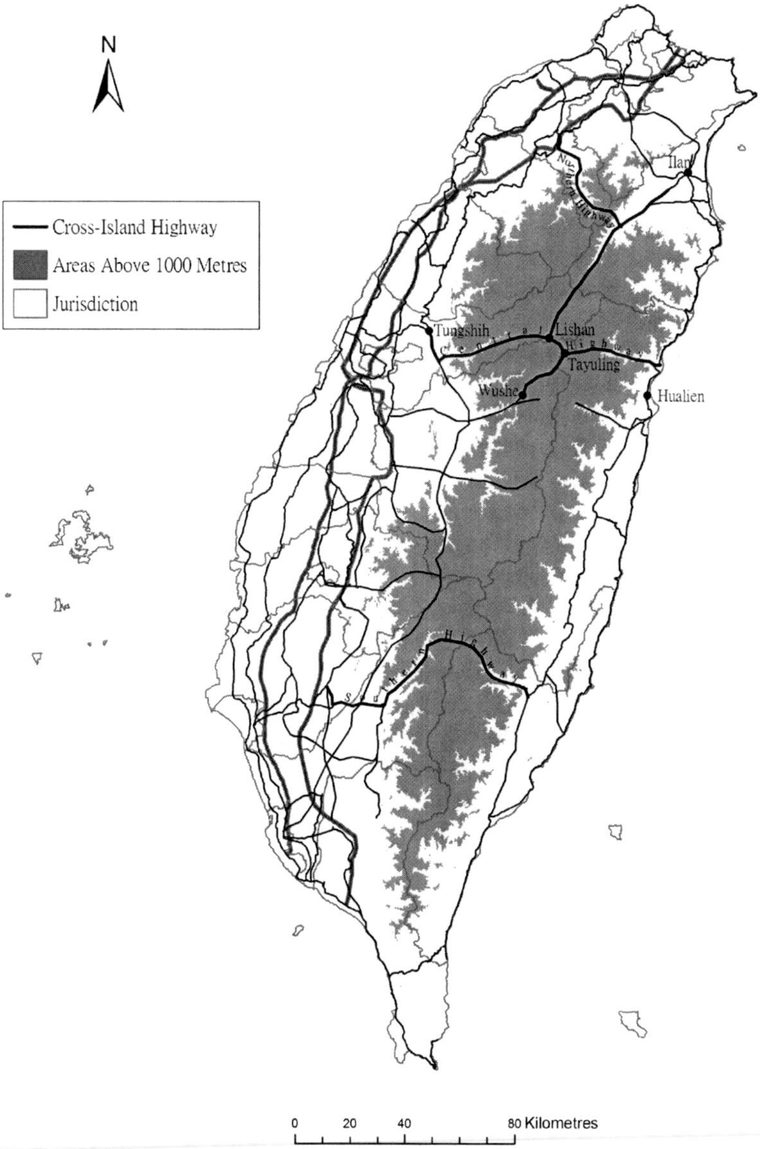

Figure 4.1 Cross-island highways (source: Ministry of Interior).

most of the 'fruits' (in the broadest sense of the word) of the effort were harvested, and here where disaster finally spelled the doom of the plan more than 40 years after the highway's construction.

One might say that it was inevitable that introduction of modern transportation access to a previously isolated territory with an attractive resource base would lead to spontaneous settlement by migrants moving in along the road corridors. It was also probably inevitable that the planned settlements built by the government for retired soldiers, without strong controls over the use of the land surrounding those settlements, would attract spontaneous settlement and eventually abuse of the land. VACRS was established by the KMT government in 1954 as a major conglomerate designed to promote the interests of the Mainlanders, particularly the military, who fled to Taiwan in 1949. VACRS had some 100 subsidiary organizations involved in all phases of the island's economy, society, and politics, and was a powerful organization at that time. The cross-island highways, as well as most other major public works projects around the island during the martial law era, were built by the Ret-Ser Engineering Agency (RSEA) of VACRS.[9]

In the upper Tachia basin, VACRS established two farms for retired servicemen (eventually, there were 14 VACRS farms in all, but 11 of them were in lower elevations). The two farms in the basin were Fushou Shan (FSS), near Lishan, and Wuling, about 25 km up the valley from Lishan. Chiang Ching-kuo personally led the team that selected the site for FSS, on a relatively gently sloping top of a mountain spur right above Lishan at the strategic junction of the planned cross-island highway and its two connecting roads to the north-east and south-west. FSS was founded in 1957, before the highway was even completed; Wuling opened in 1963. Both farms were about 500 ha in size, with more than 350 farm households each. Both were operated about the same, with carefully selected retired servicemen, chosen on the basis of seniority and military records, and run in a sort of combined collective and free-enterprise system. VACRS clearly hoped that the two farms would serve as model demonstrations of how resources in the upper slopeland zone could be developed for agricultural production, mainly for temperate fruits (especially apples, pears, and peaches) and temperate vegetables (cabbages, etc.). The government was totally unprepared, however, for how powerful the demonstration effect would be, and that its ostensibly noble plan to improve the livelihood of the aborigines would get so out of hand and end up drawing large numbers of lowlanders to the high mountain country.

Under the government's programme for development of the aboriginal reservation, which extended far beyond just the upper Tachia basin, each aborigine was allowed to stake out 0.8 ha of Aboriginal Reserved Land. Use of the land was supposed to be in accord with the suitability of the land for various purposes as determined by the government. The reality, however, was that the aborigines simply staked out squatter's rights to a piece of land, cleared the forest, and then went into the fruit growing business. An aboriginal family of four thus could theoretically have exclusive and free use of up to 3.2 ha of land (exempt from taxes for the first eight years, and then permanent ownership granted after the tenth year). With temperate fruits selling for prime prices in lowland markets, the profit potential was significant, far beyond anything the aborigines had ever experienced before. Thus, aborigines began staking out virtually all the land up and down the valley along the highway, clearing off the forest cover (whether Aboriginal Reserved Land or national forest land that supposedly was being protected from cutting), and planting fruit trees. The cultivation generally extended only as far up the mountain side as growers were willing or able to reach on foot or by simple three-wheeled vehicles developed in Taiwan for slopeland use. Transporting heavy, bulky fertilizer and equipment, and harvested produce, was difficult, so distance from the main highway mattered. Some farmers also experimented with a mini-monorail developed in Japan. Since there was no bridge across the reservoir, cultivation was restricted to the south slopes of the river valley at first. By the late 1970s, however, some cable cars were strung across the river downstream from Lishan, enabling some cultivation to open up on the north slopes, but the main cultivation area remained on the south slopes on both sides of the highway (Figure 4.2).

Many entrepreneurial aborigines took things one step further and found it still profitable and much less work to lease their land to lowlanders, many of them wealthy business people or government figures who in turn subleased the land or

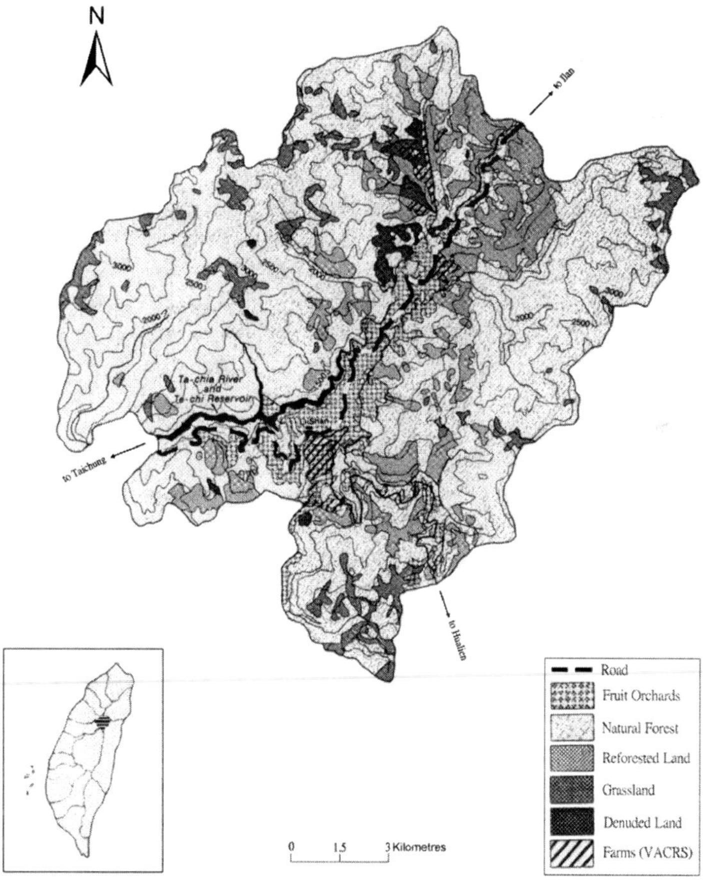

Figure 4.2 Land use in the Upper Tachia Valley, *c.*1970 (source: Williams, Chang, and Wang (1983)).

hired lowland farmers, or sometimes other aborigines, to do the work. Fruit cultivation reached about 5,000 ha (about 8 per cent of the total area in the basin) by 1980, the largest clustered area of high-elevation slopeland cultivation in Taiwan at that time. The annual income per farm family in the basin came to US$29,000 in 1981, or an estimated US$110 million for all of the 3,900 farm families operating there then. This was very good money for that time in Taiwan, especially just from farming. Most lowland farmers made far less. The end result was a complex system of absentee landlords and leasing and subleasing, all of it technically illegal, but which the government was unable or unwilling to police or stop.

The settlement and development process escalated when VACRS built a luxurious tourist hotel complex in the town of Lishan, in classic Chinese style, to promote the tourist industry and provide additional employment for servicemen.[10] Tourists eager to escape the heat and congestion of the lowlands (Taiwan's population soared to over 17 million by the end of the 1970s) flocked to Lishan, creating demand for other hotels, restaurants, and souvenir shops, in a classic story of spontaneous development. Lishan and the upper valley became one of the top tourist attractions in Taiwan, typically visited as part of a swing along the cross-island highway to also visit the famous Taroko Gorge on the east side (see Chapter 3). By 1979 more than 3,300 permanent residents lived in Lishan alone. Lishan was a boom town, leading to spontaneous settlements emerging up and down the valley in classic strip-settlement fashion, in the towns of Huanshan, Chiaying, and Sungmao. By the mid-1970s, the upper basin's total resident population was over 49,000, an astonishingly high number of people for such a fragile natural environment. Added to that could be several thousand transient tourists on a busy summer weekend or holiday, arriving in private cars or tourist buses spewing diesel fumes. Traffic jams along the highway were not uncommon.

One might say the government's ambitions for the high mountain country were too successful. Countering the benefits to the aborigines and consumers in Taiwan were the negative consequences on the environment, on the aborigines, and on the public attitude toward land and resources and how they should be developed. Only now, half a century after the cross-island highway was started, have government and public begun to come to a consensus on the fatal flaws in the plan to develop the high mountain country. Whatever the theoretical merits of the opening of the upper Tachia basin to settlement and development, the unintended consequences far outweighed the benefits in the eyes of critics.

Impact on the natural environment

The greatest environmental problem resulting from opening up the basin was, not surprisingly, soil erosion and its consequences. Taiwan's slopelands have always been prone to high rates of natural erosion, including sheet erosion, gullying, and landslides. This erosion is due to the weak rock structure, steep slopes, and high annual precipitation, especially that associated with typhoons, which typically strike the island several times each year during the summer and autumn months. A single typhoon can generate rainfall in the hundreds of millimetres in 24 hours. On-site survey in the upper Tachia Basin in the 1980s by the authors of this volume revealed the average slope of the fruit orchards was 30–35°, but slopes up to 60° were sometimes used. The government's then Mountain Agricultural Resources Development Bureau (MARDB) at that time recommended no cultivation above 28°. Hence, most of the orchards were operating on land beyond the recommended degree of steepness. Only the VACRS farms operated on permissible slopes, but that was hardly surprising in that they had staked out the best land at the very start of the process. The authors also found that soils were very shallow, averaging only 20–80 cm; whereas the FAO at that time recommended a minimum depth of 80–100 cm for proper slopeland cultivation. Most of the farmers also did not practise sound soil conservation measures (proper terracing, use of ground covers, following the contours of the land, etc.), largely ignoring the directives of the MARDB and other agencies operating in the area. In other words, an essentially unregulated, free-for-all, highly commercial agricultural system flourished in the basin.

After the Techi (Deji) Dam was completed at the entrance of the basin in 1973, the erosion problem became more visible, because of rapid siltation in the reservoir behind the dam, creating long-term risk to the generating capacity of the dam and its four sister dams downstream, which together supplied one-third of Taiwan's total hydroelectric power in the 1970s and 1980s. Moreover, there was growing concern about pollution of the river from untreated sewage flowing into the river, plus the run-off from large amounts of chemical fertilizers and insecticides used in the orchards. The Tachia River was a major source of water for over a million people living in the lower basin.

Impact on the aborigines

The cross-island highway project also had a profound impact on the aborigines. The Atayal had long been the native inhabitants of this area, and were the second largest tribal group of aborigines in Taiwan. The Japanese counted some 34,000 Atayal (also sometimes known as the Taiyal), although only a small number then lived in the upper basin. By the early 1980s, the Atayal population was about 60,000, out of a total mountain aboriginal population of just over 143,000. There were 3,171 Atayal registered as residents of Taichung County in 1979, which just about equated with the Atayal population of the upper basin. Hence, the Atayal had become a minority population in their own territory.

In principle, the ROC government, under the KMT, administered the mountain country somewhat the way the Japanese had, insofar that the territory was supposed to be restricted, with permits needed by lowlanders (or foreigners) to travel into it. The goal, supposedly, was to allow the aborigines to develop their livelihood at their own pace and preserve as much of their traditional culture as possible while bringing themselves up to 'modern' standards.[11] The reality was that aborigines had great difficulty maintaining viable local economies, with low per capita incomes and high unemployment rates, and hence young people were inclined to leave and move to the lowlands where job opportunities were greater. Maintaining native languages, dress, and customs proved very difficult also, as young people wanted to enjoy mainstream life and culture. Hence, development schemes such as that in the upper Tachia basin had potential to help the aborigines create an economic niche and possibly preserve something of their cultures. In reality, though, only some of the Atayal prospered under the new plan, mainly those lucky enough to stake out the good land early in the process, or with the luck and skills to develop lucrative leasing arrangements. Most of the Atayal, however, were outclassed by clever lowlanders who gradually came to dominate the basin. Understandably, the ROC government, through VACRS, MARDB, and other agencies, was very sensitive to anyone probing too deeply into the situation in the basin (not to mention questioning the wisdom of the cross-island highway, or the Techi Dam and other structures), as these two authors found out in their field work in the 1970s and 1980s. Was this just a noble desire to protect the aborigines from exploitation, or determination by the KMT to shroud its activities in the basin under a cloud of secrecy and keep prying eyes out? Opinions on this question varied, depending on to whom one talked at the time.

Impact on public attitudes toward land and resources

One of the most insidious consequences was the utter disregard for laws and regulations designed to protect the environment. The reluctance or inability of the government (in reality, a combination of both factors) to enforce the laws and regulations encouraged an attitude of indifference to public welfare (i.e., the greater good of the people of Taiwan as a whole) for the sake of private gain. As a result, illegal cultivation of slopelands spread to other areas of the mountain region, particularly along the other cross-island highways and forest roads that have provided additional access to the mountain slopelands. (Today, there are many hundreds of miles of logging and other roads into the mountains that, however primitive some may be, provide access into the interior.) Since much of the activity on slopelands was illegal, it is impossible to give even an estimate of the total area or number of individuals involved at that time.

To be fair, the government did make some efforts to rein in unplanned and illegal activity in the mountains, mainly through studies carried out at various times, and recommendations by academics and government officials as to proper rules and regulations that ought to be followed. In 1967 the Lishan Administration Office was established as the principal government body directly overseeing development of the area, but had very limited success. Thus, in 1979 MARDB was given authority over registering and supervising aboriginal land claims. In 1980 the government came forth with three options applying to land over 28° in slope that was being cultivated in the upper basin: (1) land on which erosion or landslides were physically evident would be taken back from the farmers immediately, regardless of their status (aborigine, lowlander, or retired serviceman); (2) land on which soil erosion was a potential problem would be taken back after one year; (3) land on which no soil erosion was likely would be removed from cultivation after the existing fruit trees died out (normally 8–20 years later). None of these actions were implemented, however. More than 25 years later, the government appears finally ready to impose similar hard-nosed measures (see page 150).

The MARDB was restructured as the Bureau of Water and Soil Conservation (BWSC) in the early 1990s, but no other agency took over some of the former tasks of the MARDB, although the Soil Conservation Law passed in the early 1990s did contain some provisions previously proposed by the MARDB on slopeland farming. With passage of the EIA Law in January 1994, supposedly any new construction and development in the upper Tachia basin (or anywhere else, for that matter) had to pass the EIA process first. However, this obviously impacted construction and more formal projects, not the illicit activities of farmers (aborigines or otherwise). Hence, private cultivation generally went on without interruption right up to the present time, although more than half the farmers now are said to follow 'good' soil conservation practices. Illegal cultivation still continues, though, especially on steep slopes.

The data in Table 4.1 show some facets of the situation as of 2005. According to this government data, there were a total of 4,521 aborigines living on land above 1,500 metres, and another 68,000 living between 500–1,500 metres. In other words, the vast majority (some 84 per cent) of aborigines are now located in the lower slopeland zone (100–500 metres). Only about 4 per cent, or almost 11,000, of the 260,000 ha of land reserved for aborigines lies above 1,500 metres, while the majority (55 per cent) is in the intermediate zone, and the remainder (41 per cent) in the lower zone. Thus, relatively small numbers of aborigines and reserved land are found in the highest elevation zone. Nonetheless, the upper Tachia basin, and its cultivated fields, lies in the transition area between the two slopeland zones, from about 1,200 metres and higher. Also revealing are the data for all land under cultivation in the three zones by all parties concerned (aborigines, Taiwanese, Mainlanders). Officially, only about 2,400 ha were being cultivated at the highest level in 2005, but a significant 73,000 ha were being used in the middle zone; while the bulk of cultivated land (more than 78 per cent) was in the lower zone. Yet, the production value and value per ha in the highest elevation zone were quite high, suggesting the rewards to be earned from growing crops at that elevation.

The various crops planted in the three slopeland zones are detailed in Table 4.2. In the highest area (over 1,500 m) various fruits and dry (non-irrigated) crops dominate, but the total area (officially at least) is not large. While the main cultivation takes place in the lowest zone (100–500 m), the middle zone contains a surprisingly large amount of cultivation, with fruits of various kinds very important, especially peaches, plums, apples, and others. Indeed, fruits of all kinds account for fully 42 per cent of total cultivated land in the three slopeland zones, clear indication of the extreme importance of fruit production to the future of agriculture in Taiwan (see below). The persistence (and problem) of betel nut cultivation is also evident, accounting for just under 7 per cent of all

Table 4.1 Population and agriculture in Taiwan's slopelands, 2005

Slopeland elevation level (m)	*Aboriginal reserved land (ha)*	*%*	*Aboriginal resident population*	*%*	*Land under cultivation by all parties[1] (ha)*	*%*	*Production value (NT$million)*	*Production value per ha (NT$million)*
>1,500	10,887	4	4,521	1	2,388	0.7	3,400	1.424
500–1,500	141,592	55	68,000	15.4	73,138	20.8	19,200	0.262
100–500	107,521	41	367,479	83.6	276,025	78.5	22,300	0.080
Total	260,000	100	440,000	100	351,551	100	44,900	

Source: Unpublished data from Council for Economic Planning and Development, 2006.

Note

1 Includes both aborigines and lowlanders (Taiwanese and Mainlanders).

Table 4.2 Crop planted area in Taiwan's slopelands, 2004

Crop	*Area >1,500 m*	*Area 500–1,500 m*	*Area 100–500 m*	*Total area*
Hay	–	162.5	1,900.0	2,062.5
Tea	25.0	4,162.5	14,087.5	18,275.0
Sugar	–	250.0	17,037.5	17,287.5
Vegetables	50.0	2,200.0	912.5	3,162.5
Other dry crops	950.0	13,975.0	46,512.5	61,437.5
Bananas	–	162.5	3,237.5	3,400.0
Pineapples	–	12.5	2,925.0	2,937.5
Oranges	212.5	4,400.0	21,887.5	26,500.0
Peaches, plums	187.5	15,775.0	7,275.0	23,237.5
Apples, pears, peaches	–	8,087.5	2,925.0	11,012.5
Betel nuts	175.0	6,362.5	16,537.5	23,075.0
Other fruits	375.0	9,200.0	72,675.0	82,250.0
Other crops	362.5	6,262.6	13,175.0	19,800.0
Paddy (rice)	50.0	2,125.0	54,937.5	57,112.5
Total area	2,387.5	73,137.5	276,025.0	351,550.0

Source: Unpublished CEPD data.

Note
All figures in Ha.

cultivation in the slopelands (see Chapter 3). Overall, the slopeland cultivated area accounted for over 40 per cent of Taiwan's total cultivated area in 2004, presenting a significant challenge to government efforts to control soil erosion and related problems.

Unfortunately, by the end of the 1990s and into the first decade of the twenty-first century, continued human impact on the mountain slopelands was coming face to face with increasingly intense natural disasters, almost as if nature were trying to punish Taiwan and its people for their reckless behaviour in the mountains.[12]

Large sections of the Central Cross-Island Highway were devastated when a major earthquake struck central Taiwan on 21 September 1999, one of the most serious earthquakes of the twentieth century to affect the island (Figure 4.3). A follow-up quake on 11 June 2000 finished the job of wiping out the section of road between Techi and Kukuan downstream. Rebuilding the stretch of highway destroyed in those quakes required five years and millions of dollars. Just as the highway was to have its grand reopening in July 2004, Typhoon Mindulle hit Taiwan, one of the most powerful storms ever to rake the island, with a level of rainfall that occurs once in 200 years.[13] During the three days that Mindulle lingered over Taiwan, more than 1,000 mm of rainfall was recorded in many parts of central and southern Taiwan, and one station recorded 2,000 mm, almost the normal annual rainfall of Taiwan.[14] Mindulle triggered huge landslides and mudflows throughout much of the island, including between Kukuan and the Techi Reservoir in the upper Tachia valley (Figure 4.3). The slides knocked out the same vulnerable stretch of highway destroyed in the 1999 quake. This is the main route for farmers in Lishan and Hoping townships to transport their fruits and vegetables to market, so the hardship imposed by several years of closure has been great. To get their produce to market, the farmers in the upper basin have had to transport their fruits and vegetables via the spur roads, north-east to Ilan or south-west to Wushe and beyond, adding to shipping costs and hence cutting into profits.

Economic losses from the storm were estimated at about US$400 million. Homes and other structures were buried, bridges and roads washed away, farmland and orchards wiped out, and more than 2.4 million fowl and livestock drowned. Officially, 29 people were killed. Six state-run power plants along the Tachia River were inundated and damaged by floodwaters. As illustration of the volume of water involved, a suspension bridge at the Kukuan hydroelectric plant was once 22 m higher than the surface of the river. After Mindulle arrived, the flood waters were 4 m higher than the bridge and washed it away.[15] The Kukuan plant was supposed to be rebuilt and reopen in 2007. However, subsequent typhoons finished destroying the plant, which has been abandoned. Mindulle thus was the most devastating typhoon to hit Taiwan in recent years.[16]

Adding insult to injury, Typhoon Aere hit the island in early August 2004, bringing on further destructive mudslides, destroying a number of aboriginal villages in Hsinchu County, and wreaking further major damage on the agricultural sector.[17] Another consequence of Aere was to trigger a massive influx of mud into the Shihmen Reservoir in Taoyuan County. Shihmen was already one of the most turbid of Taiwan's reservoirs, and this latest assault made the water too muddy for plants to process the water for public consumption, resulting in serious interruption of supplies and water shortages for people in the Taoyuan area.[18] A study out of National Taiwan University noted that the number of rivers (counting all minor tributaries as well as main streams) in Taiwan prone to mudslides and rockslides had surged to 1,420 since the September 1999 earthquake, of which 655 of the most 'dangerous' rivers were in central Taiwan. This was in contrast to just 485 such rivers throughout the island in 1996, with 134 in the central region.[19] What had changed was denudation of slopes and increased propensity for erosion.

The one-two punch in the summer of 2004 triggered a paroxysm of recriminations and political infighting in Taiwan, and released pent-up frustrations by environmentalists and others who felt aggrieved over what they saw as years of blind development policies and neglect of nature. This was hardly a new situation in Taiwan, although to be sure the devastation in 2004 and recent years was greater than usual. But, for example, back in 1996 Typhoon Herb, then called the worst storm to hit Taiwan in 30 years (although Mindulle in 2004 was far worse), exposed once more the government's inadequacies in disaster-control measures, most glaringly the failure to enforce restrictions on land use, especially in mountainous and hilly areas.[20] For example, the second Central Cross-Island Highway, which had been rerouted but still went through fragile areas in Hsinyi Township in northern Nantou County, was finished in the early 1990s. But then Typhoon Herb triggered many landslides along that highway, and the problem has recurred year after year since then. In 2004 and 2005, a

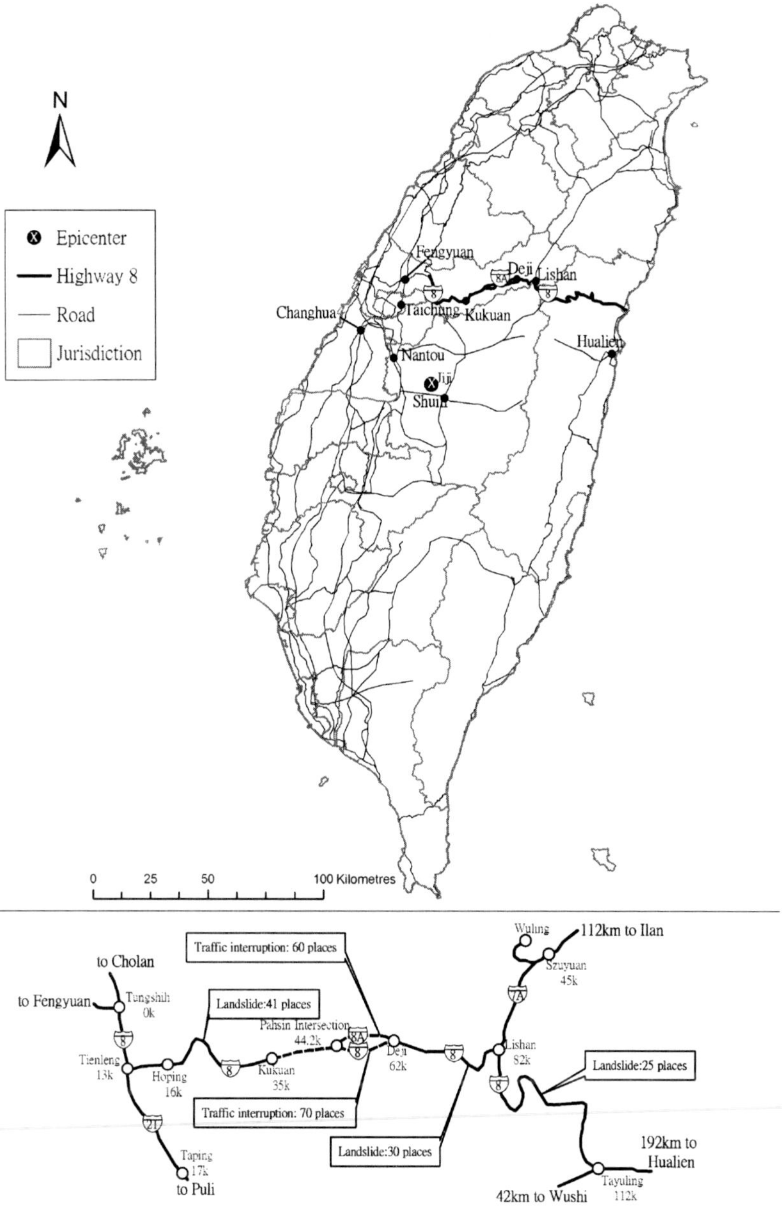

Figure 4.3 Central Cross-Island Highway and the 1999 earthquake (source: Ministry of Interior).

rising chorus demanded that real change be instituted this time, not just bland pronouncements of good intentions and yet more studies that would sit on shelves unread and unheeded. It was time for action.[21]

Healing the wounds: the national land recovery plan

Unfortunately, ethnic relations and politics got mixed into the aftermath of the 2004 disasters. In the search for someone to blame, some members of Taiwan's government claimed the aborigines had overdeveloped the slopelands in the disaster region, which lies in an aboriginal reservation. But others countered that efforts by non-indigenous farmers and businessmen to usurp aboriginal land rights was the main cause of overdevelopment of the land that had already been weakened by the 1999 earthquake and subsequent typhoons.[22] A storm of controversy and counter-charges raged on through 2004 and 2005 as Taiwan struggled to recover from the ravages of Mindulle and Aere, and to prepare for future disasters. A significant change in government and public attitude began to emerge. Now the government and planners, urged on by environmentalists, started seriously thinking about not rebuilding the highway, at least for the time being, while the area's natural environment is allowed to recuperate and planners have a chance to carefully study options.[23]

By late 2004 a National Land Recovery Plan was gestating in the legislative process. The key philosophical thinking in the plan was to:[24]

1 Respect and adapt to nature instead of believing that humanity can dominate nature.
2 Consider policies from the standpoint of sustainability instead of only the present.
3 Foster 'green' ecologically friendly economic development instead of refusing to consider environmental and ecological resource costs.
4 Utilize resources based on the distinct environmental character, regulate exploitation and recovery measures instead of engaging in unrestricted development.
5 Carry out regional environmental regulation from an integrated and nature-friendly stance instead of adopting scattered and uncoordinated measures carried out by different and competing agencies.
6 Stress 'light and soft' ecological engineering to 'guide' and 'channel' water flows or other pressures in project engineering instead of stressing 'hard' engineering methods aimed at 'blocking' or 'obstructing' such flows.
7 Stress returning land hit by disaster to nature and rely on management more than control in dealing with natural disasters, instead of hardware engineering that aims to repair the damage to man-made installations.

In short, the DPP government was trying to make the transition to a more ecologically friendly, sustainable development path for Taiwan's future, which it had long claimed was its goal and part of the party platform (see Chapter 3).

More specifically, the recovery plan consists of three parts: a National Land Recovery Act, a Coastal Zone Management Act, and a related National Land Recovery Strategy and Action Plan (most important of the three). The documents were approved by the DPP Cabinet in January 2005, and approved by the Executive Yuan on 20 January. The statute, however, was still tied up in the Legislature as of early 2007 because of Pan Blue/Pan Green gridlock. Nonetheless, the government can carry out some of the plan's provisions, even though the plan is not law yet. For example, if a typhoon strikes and affects a village over 1,500 m elevation and the road washes out, the government will not invest money to repair the road, in line with the plan's provisions.

The plan would divide Taiwan into three areas of high, medium, and low zones of land sensitivity and protection, as follows:[25]

1 High protection zones: Areas in mountains above 1,500 m elevation, uninhabited islands, and some coastal areas would be designated as high protection areas. All new development in these areas would be forbidden. Existing developments and structures would be removed, except for indigenous peoples' (aboriginal) villages, cultural relics or installations needed for ecological protection, forestry, defence, or conservation uses. Aborigines would be allowed to continue cultivation for self-sufficiency and be granted special rights for ecological tourism and related services, but all large-scale commercial agriculture, such as fruits, tea, and vegetables would be banned.
2 Medium protection zones: Areas in mountains between 500 and 1,500 m would be designated for this category. New agricultural and other developments would be banned, but existing legal operations in agriculture or other land or tourism developments would be allowed to continue. All illegal operations or developments would be stopped.
3 Low protection zones: Areas between 100 and 500 m elevation and most coastal areas would be classified in this zone. The operating principle here would be 'sustainable development', and central and local government land use plans and licensing would be subject to regular review and approval by the central government.

In addition, the draft law would designate eight 'national land recovery promotion zones' for which special recovery plans would be drafted and implemented. The eight categories include: (1) areas of high risk of land or mudslides in river catchment zones due to past over-exploitation; (2) severe rock collapses; (3) concentrations of excessive land use; (4) severe land slippage; (5) rivers with ecological degradation or concern of flood safety; (6) areas of severe ecological damage; (7) illegally occupied or developed land; and (8) other areas of severe damage to national land. Obviously, the Upper Tachia Valley would be one of these special land recovery promotion zones. Once approved, the statute would require illegal farms or structures to be removed within five years. The government would exercise the right of eminent domain to take over land being used legally with compensation. Mass tourism in sensitive areas would have to end, but 'ecological tourism' with controls on volume and intensity of activity would be allowed. In addition to not rebuilding the Central Cross-Island Highway for the immediate future, the plan also calls for not repairing roads in areas with fewer than 30 households.

In some respects, these new regulations and plans appear to be a reiteration and expansion of the directives laid out more than 25 years ago that were never effectively enforced at that time (see page 144).

To help carry out this unprecedented reordering of land use, an official 'national land recovery fund' would be established with an allocation of at least NT$100 billion (US$4 billion) during the coming decade to cover the costs of buying land from currently legal operations in protected areas, plus other subsidies, assistance, relocation, and living expenses for persons affected. In other words, the government would do all it could to help persons involved in current legal operations to re-establish themselves in new locations and possibly new occupations.[26]

There is also a strong push now from environmentalists and others to employ what is called 'ecotechnology' or 'ecologically friendly engineering' to rebuild and restore the mountain environment and public structures that must be rebuilt. 'Ecological engineering' was a term coined in 1962 by an American engineer, Howard Odum, and refers to the use of environmentally friendly technologies that are efficient, clean, and adapted to local conditions.[27] As just one example of this new approach, after the 1999 earthquake, in rebuilding dikes along one of the rivers, engineers built wood piles, boulder embankments, and rearranged rocks already in the river to help control current speed, rather than the more traditional engineering approach of pouring more concrete.[28] The government even held an International Symposium on Ecological Engineering in November 2006, to get input from foreign experts on how to employ sustainable development using engineering projects that minimize damage to ecosystems.[29]

To help eventually enforce the plan's provisions, including stopping illegal logging and poaching, in July 2004 the government launched a new special police force, dubbed the 'forest cops'. Operating under the authority of the Council of Agriculture and the Ministry of the Interior, the task force was to operate for one year on a trial basis. These 'cops' were authorized to help Taiwan's existing force of forest rangers, who are not authorized to make arrests in cases of arson, poaching, illegal logging, or dumping. Thus, the forest cops, who were to be informal aides to the rangers, would help provide manpower in the government's efforts to crack down on illegal activity in the mountains. It is a daunting task, and perhaps understaffed and underfunded, but at least it was a start. The force would operate in seven counties in the south, helping just 88 forest rangers patrol over 187,000 ha of forest, or roughly 2,100 ha for each ranger. This area intersects with the aboriginal land, and aborigines were to be included in the force, in an effort to win over, rather than alienate, them. Cooperation from the aborigines obviously will be essential to make such an operation a success. Eventually, the government hoped that 80 per cent of the force would be aboriginal.[30]

To environmentalists, the provisions and regulations in this new land use act seem sensible and very much in line with their understanding of 'sustainable development'. Many aborigines do not share that perspective, however. They see themselves facing relocation, drastic changes in their traditional lifestyles, and loss of incomes. Those living in the Tachia River valley were especially upset about plans not to rebuild the destroyed section of the Central Cross-Island Highway. Clear evidence of this was visible in late 2006 when the authors travelled through the upper Tachia valley and saw numerous home-made signs and banners posted in various places. These signs protested the government's plans and proclaimed the 'right' of the farmers in the basin to earn a living from farming. The aborigines see the roads as their lifelines to the outside world, and are not going to acquiesce quietly to the recovery plan without solid guarantees of assistance from the government.[31] The Taiwan Aborigine Farming Association led a demonstration in March 2006 in front of the Legislature in Taipei to protest what they perceive as discrimination against the aborigines and infringement on their 'rights', urging the government to rethink the National Plan.[32]

This aboriginal resistance to government plans, however well-intentioned those plans may be, is very much a consequence of the post-martial law reform era, which has seen a reawakening of ethnic identity among many of the aborigines, who are becoming more and more assertive about their perceived 'rights', even if those rights sometimes go against sound environmental practices. Thus, the government is being forced to practise increased sensitivity toward aboriginal viewpoints and interests in a way that would have been unthinkable in the martial law/KMT era.

A Council for Indigenous Peoples now exists under the Executive Yuan, with representatives from the CEPD, Ministry of Interior, Council of Agriculture, and other agencies. The proposed Ministry of Environment and Resources (MER) (see Chapter 3) would be in charge, if and when that ministry eventually comes into existence. As things now stand, the aborigines will have true autonomy. An Aboriginal Autonomy Law has already been passed. Villages above 1,500 m will be allowed to stay. At lower elevations, aboriginal villages of 50 or more population will be allowed to vote among themselves whether or not to stay in their mountain reservation. Current thinking among experts in Taiwan is that some of the 12 official tribes may vote to subdivide into separate groups on the basis of the new autonomy law.[33] The tribal areas, however, will still be subject to federal law and to regulations on land use. In other words, the current tribal boundaries in Taiwan, and the boundaries of the Aboriginal Reserved Land, are in a state of flux and very likely to change in the near future as the aborigines sort out their views and positions, and as the government adapts to and tries to work in a cooperative fashion with the aborigines, in ways that were never required in the past.

The aborigines and Taiwan's environmental future

Taiwan's future, especially in the environmental realm, is increasingly intertwined with the aborigines and their push for 'direct democracy' and economic rights. In February 2005, the administration put into effect a new Aboriginal Basic Act, which states that the government must safeguard the status of aborigines and work toward providing self-rule of each tribe, as the best way to protect the rights and heritage of Taiwan's indigenous groups. This new law was fulfilment of a pledge that Chen Shui-bian made in his 1999 campaign for the Presidency, to establish a new, equal partnership between the Han majority population and the indigenous minority. This new act followed on the heels of a Draft Law on Indigenous Autonomous Areas approved by the Executive Yuan in June 2003 (which was slated to come before the legislature for a vote in the fall of 2006). It is the dawn of a new era for the aborigines, and by consequence for Taiwan as a whole.

Politically, this new era is illustrated by the experience of the Truku (a subgroup of the Atayal tribe) and the Saisiyat. The Truku were formally recognized in 2004 as the island's 12th distinct aboriginal group.[1] The Truku have led the way in pushing for self-government, brought about in part by the establishment of Taroko National Park in 1986 (see Chapter 3) and government efforts to develop the nature conservation and cultural preservation of the park without consulting the Truku, the native inhabitants of the area, during the martial law era. However, after martial law was abolished in 1987, direct democracy has been in progress, and complex procedures have to be followed to make self-rule a reality, including passage of the draft law on autonomous areas noted above. Many aborigines, such as the Saisiyat, also want to see an end to the provision in the ROC Constitution that divides all aborigines into two classes: 'plains aborigines' and 'mountain aborigines', which are regarded not only as simplistic and inaccurate labels, but also tend to divide the aborigines politically and reduce their power. Hence, this political process is still unfolding.

The other dimension of the 'indigenous rights movement' is the quest for economic independence. Long relegated to the bottom of the socioeconomic ladder in Taiwan, the aborigines are no longer content to stay there. Numerous, but often misguided, efforts by the government in the past to provide an economic base for aboriginal villages are now beginning to be supplanted by new approaches. In the 1970s, for example, the government promoted cultivation of cabbage and other temperate vegetables, as well as tree fruits, building on the experience of the VACRS farms for retired servicemen (see Chapter 4). Unfortunately, those experiments often led to overexploitation of fragile soils, inducing slopeland instability, ever-increasing fertilizer use causing pollution problems, decreasing yields, and now competition brought on by membership in the WTO. 'Tribal eco-tourism' is an alternative strategy being experimented with today. This involves entire villages promoting ecological and cultural resources to market themselves as tourist destinations, in a form of what is commonly called 'eco-tourism' or 'green tourism'. Traditional foods are served and costumes worn, and traditional festivals and other customs observed for the benefit of the tourists. While this approach holds some promise for helping preserve cultural traditions and the cohesiveness of the village, there are definite drawbacks that have to be addressed. Villages of just 100 residents, say, can be overwhelmed by as many as 300 tourists in one day. Food must be brought in from the lowlands, and the village women have to prepare meals. Travel agencies bring the lowland tourists to the area, but take a substantial cut. Environmental destruction, rising land prices, disputes about distribution of profits, and other problems arise. The aborigines have much to learn, and their options are actually quite limited. But they are not content to be passive spectators anymore.

Sources: J. Tsai, 'Aboriginal groups push for greater autonomy', *Taiwan Journal*, 30 June 2006, p. 2; J. Tsai, 'Saisiyat people launch referendum initiative', *Taiwan Journal*, 28 April 2006, p. 2; C. Lee, 'Economy or identity: the choice for aboriginal development', *Taiwan Journal*, 30 June 2006, p. 8. Also see: Chang and Liang (2005), Chang and Lin (2006), Tsai *et al.* (2006b).

1 As this book was going to press, a 13th aboriginal group was recognized by the government. See Note 33 of this chapter.

This is a highly sensitive area of public policy in Taiwan right now, made more complex by the ongoing political standoff between the Pan-Blue and Pan-Green camps, with the aborigines sometimes caught in the middle of the cross-fire. For example, an NGO known as the Taiwan Aboriginal Land Development Rights Alliance (established in the 1980s), made up of non-indigenous business people, petitioned the Executive Yuan to open aboriginal reservations to the real estate market. That has been done, so the aborigines now own much of the mountain land.[34] Adding fuel to the fire, Vice President Lu was caught making injudicious statements to the effect that the aborigines were not the true 'native people' of Taiwan, but that everyone in Taiwan had migrated there at some time in history, a rather bald effort to diminish the legitimacy of aboriginal land claims.[35] Needless to say, actions and statements like these from the majority lowlanders of the island do not help win over support of the aborigines for the government's policies and programmes in the mountain country.

Also caught in the middle is the environment. As one example of the government's difficulties in dealing with now assertive aborigines, when the government in recent years has tried to evict illegal cultivators, the aborigines have retaliated on occasion by burning down the forest, forcing the government to back down. Thus, the government rarely tries to evict them now, but rather tries to help the aborigines develop better land-use practices, even if they are illegally using the land. This compromise position, however, effectively legalizes their use of the land, and raises troubling questions for the future, when the provisions of the new Land Recovery Act must be enforced. The potential for serious conflict in the slopelands exists. Much will depend on how the government goes about trying to put the Recovery Act into operation. Many lowlanders wonder whether the law will be enforced or not. For example, the government has hopes of resettling many of the aborigines in new residential communities to be built on suitable land owned by the state-run Taiwan Sugar Corporation, which has lots of land inherited from the Japanese era but which now grows relatively little sugar.[36] That relocation process is not likely to be easy.

However, relocation of aborigines is hardly something new in Taiwan. Indeed, aboriginal groups have had to shift their locations repeatedly throughout Taiwan's several centuries of settlement and development (see Chapter 1), in response to migration by Hoklo, Hakka, Japanese, and then Mainlanders, and in response to government policies. As Taiban Sasala, Chairman of the Kaohsiung City Government's Commission of Indigenous Affairs, put it so well in a newspaper editorial, the connection between aborigines and their land is deeply tied to their history, culture, and collective memory. Hence, it is no small affair to force them to relocate to an environment that may be quite different from what they are used to, and making it that much harder to maintain traditional culture and tribal members together. Sasala reminds readers that Article 10 of the UN Draft Declaration on the Rights of Indigenous Peoples states that 'Indigenous peoples shall not be forced from their lands or territories … No relocation shall take place without the free and informed consent of the indigenous peoples concerned and after agreement on just and fair compensation and, where possible, with the option of return.'[37] These are noble principles that are routinely violated by states dealing with their own indigenous peoples. In other words, Article 10 presents major challenges to governments around the world.

The list of grievances from Taiwan's aborigines is long. For example, with the construction of the Shihmen Dam, Atayal people from the village of Kalaso were forced to relocate in 1960 and are reported to have lived a difficult life since then. The establishment of the various national parks and conservation areas in the last 20 years are said to have cost the Truku, Bunun, Paiwan, and Rukai tribes some of their hunting grounds and sacred lands.[38] Countless aborigines have lost their land and natural resources from various restrictions and policies over the years. In effect, two ways of life are in conflict in this process, and the odds are in favour of the majority Taiwanese. Yet, somehow the government has to find a way to protect the fragile mountain environment while helping the aborigines preserve their traditions. These may be inherently contradictory goals, but the effort has to be made.[39]

As for the VACRS farms, the two in the upper Tachia basin (FSS and Wuling), as well as Chingching farm, located in Nantou County on the south spur of the Central Cross-Island Highway on the way to Wushe, will eventually be required to shut down (at least in terms of cultivation), in compliance with the new Recovery Plan.[40] The other VACRS farms, which are in lower slopeland zones, will still cultivate fruits and vegetables, but those are no longer their main source of income.[41] Instead, the farms are gradually turning to eco-farming, a form of tourism where visitors (mostly lowland urban Taiwanese) can get a brief taste of farm life (see page 159). Within the upper basin, because of the closure of the cross-island highway, and the growing tide of support for restricting slopeland cultivation, the area under cultivation has definitely decreased in recent years, but there are no hard data available to compare with the situation found by these authors in their 1983 study. Likewise, the upper basin's precise population today also is not known, but definitely has fallen from the peak of nearly 55,000 in the late 1970s and 1980s. Simple observation in the autumn of 2006, however, still revealed substantial and widespread cultivation throughout the basin, and innumerable settlements and farmhouses scattered everywhere. If anything, the pattern seemed more dispersed and less concentrated in Lishan than was true in the past. There is no question that the disruption of the cross-island highway has severely impacted the whole economy of the upper basin, for agriculture as well as tourism. While tourists can still get to the upper basin, albeit by more roundabout ways from the north-east and east, and without being able to make the complete circuit across the mountains, the volume of tourism is definitely down, and those who depend on tourism have suffered.[42] The cross-island highway was like the main artery linking one side of Taiwan to the other. Now, if the government and environmentalists have their way, there will be no bypass surgery. Besides, Taiwan's admission to WTO has changed the whole economic environment for agriculture, with imports now much more competitive against fruits and vegetables grown on the island, whether in the mountains or the lowlands (see below). Thus, the profit motive for high altitude fruits and vegetables is diminished, and it may be just as well in light of the environmental problems in the slopelands.

In conclusion, thus, profound changes may be in store for Taiwan's mountain country if all of these provisions and plans are effectively carried out, and all the parties involved fully cooperate. That is a big 'if' in Taiwan's turbulent political and economic environment.

Land use issues in the lowlands in the WTO era

While the slopelands have received an extraordinary degree of attention in recent years, because of the environmental conditions and problems discussed above, the rest of Taiwan, i.e., the lowland plains, is by no means trouble-free from an environmental and land use perspective. As was noted in Chapter 1, it was agriculture that laid the foundation for Taiwan's post-1950 economic miracle, in an agricultural system based on very small, privately owned, family farms. But gradually, over the years, those small farms, averaging less than a hectare in size, became less and less viable as economic enterprises if the farmer tried to rely entirely upon crop production for a livelihood. As farm incomes began to slip in the 1960s, the government had no choice but to gradually institute price supports, tariffs against imported agricultural products, and other measures to prop up the farm system. Most farmers also had to work part-time outside the farm in industry or other activities to make ends meet. As everyone knows, farming is much more than just an economic activity. It is a way of life and part of the heritage of Taiwan. Moreover, food security, not having to rely excessively on imports for the basic necessities of life, is seen as an important issue in Taiwan, because of the uncertainty of adequate supplies always being available abroad. Hence, Taiwan developed an agricultural system very similar to that of Japan and South Korea, a system that could never survive on its own without government subsidies if forced to compete in a global free trade environment.

Over the years, young people especially lost interest in farming, drawn by higher wages and job opportunities in the urban/industrial/service economy. Farm work is still hard and low paying. Hence, the median age of farmers is steadily rising. Those who do stay in farming often find they must work part-time in non-farm occupations to make ends meet. Indeed, only a minority of farmers today support themselves entirely from farming. This is a dilemma that faces most of the Asian states with tight agricultural resources and high population density, such as Japan, South Korea, China, and others.

WTO and its impact

Then, along came the World Trade Organization (WTO), following on decades of negotiations dedicated to gradually lowering tariff barriers between nations and promoting a global free trading system that ostensibly would benefit all nations. Taiwan joined the WTO on 1 January 2002, and that date thus became a watershed point for not only the island's agricultural system but for the entire economy.[43] WTO membership would entail fundamental changes in all aspects of Taiwan's economy, with likely negative impacts on some sectors, but others benefiting, and in the long run the promise of the whole economy moving forward, with Taiwan having greater access to some 143 foreign markets.[44] The immediate beneficiaries were consumers, who enjoyed lower tariffs and hence lower prices on nearly 4,500 products.

For agriculture, however, the immediate prospects looked menacing, given the high level of subsidization and support provided by the government. WTO membership required that support to be gradually whittled down. In 2001, average nominal tariff rates on agricultural and industrial products combined were 8.2 per cent. This was to decrease to 7.2 per cent in 2002, and by 2005 the rates had to drop to 5.54 per cent. Average tariff rates on agricultural products alone were to be reduced by almost a quarter, from 20.02 per cent to 15.21 per cent (tariffs on industrial products were to fall by almost a third, from 6.03 per cent to 4.15 per cent). By the end of 2007, agricultural tariffs were supposed to be at 12.9 per cent.[45] In other words, a gradual lowering of the tariff wall is inevitable now.

Moreover, of 41 previously banned agricultural products, 18 could now be imported (including duck meat, cuttlefish, sweetened condensed milk, potatoes, lichees, papayas, pomegranates, fresh longan, citrus fruit, and others). Another 22 items could be imported under a quota system (including pork bellies, chicken, various forms of milk, mackerel, sardines, peanuts, garlic, red beans, dried mushrooms, betel nuts, bananas, pineapples, mangoes, grapefruits, persimmons, sugar, coconuts, and others). Rice and rice products were to be subject to a quota of 144,000 metric tons in 2002, about 8 per cent of the island's annual rice consumption. This change was predicted to have a greater effect on farmers than any of the other agricultural imports now being allowed in. These changes, however, did not apply to trade with the PRC, as that is governed by different regulations, reflecting the unique political relationship between Taiwan and the PRC (see other chapters). Predictions were, though, that China would begin to supply ever-increasing amounts of Taiwan's food.[46]

For agriculture as a whole, production value was predicted to fall 6.8 per cent in 2002 from its 2000 level of US$10.56 billion, and decrease another $400 million by 2004. In 2000, agricultural production contributed just 2.1 per cent of GDP, and agricultural exports less than 3 per cent of the island's total exports. Decreases were also predicted in the labour force for agriculture, which totalled 718,000 people in farming, fisheries, and forestry at the start of 2002, or 7.6 per cent of total employment (not all of these were full-time farmers, however). Thus, the bottom line is that at least the near-term future for agriculture in Taiwan did not look good.

The future of agriculture in Taiwan

So where does the future of agriculture lie, and how is that future likely to impact the environment in Taiwan? As officials in the Council of Agriculture point out, the consistent strengths of the island's agricultural system lie in technological innovation, high quality human resources, and well-established networks.[47] Starting in 2001, the government liberalized sales of farmland. Although land classified as farmland must remain in agriculture, the new policy aims to increase investment in agriculture and to consolidate fragmented fields into larger holdings in order to increase productivity. Thus, larger farm size is one strategy for adapting to the new WTO system. In the process, the government hopes that more farmers will leave farming and go into other sectors of the economy. In other words, the proportion of the labour force in agriculture is still regarded as too high in relation to Taiwan's level of economic development. Another strategy is to get some farms to convert to so-called 'holiday farms', or 'eco-farms' as noted in the previous section on slopeland farming. These are farms that open up their facilities to short-term visitors and give them a sample of farm life and the opportunity to buy farm products and souvenirs. The COA is also encouraging townships to take the process one step further and develop theme parks that include an agricultural component. By 2002 there were reported to be 46 agricultural parks, 21 'recreational farms' (i.e., holiday farms), and more than 100 tourist orchards (pick-your-own, already widespread in the

US and elsewhere). In 2006, the COA launched what it called the 'New Agriculture Movement', which aims to integrate local agriculture with production, lifestyle and ecology. A contest was started for rural and fishing villages wishing to compete for modest cash prizes and honours from the government for redesigning their villages to promote traditional industries. The programme is closely tied in with the movement toward agricultural tourism. The first ten winners were announced in early 2007, and the programme holds promise as providing support for villages to survive and even thrive in the age of WTO.[48]

The government is also encouraging farmers to develop 'niche' markets for high-quality produce that can achieve brand-name recognition outside of Taiwan, in such products as orchids, subtropical and tropical fruit (e.g. star fruit, mangoes, wax apples, pineapple), and fish farming. Developing these niche markets requires strengthening the 268 existing grassroots organizations known as farmers' and fishermen's associations, which have already begun to establish brand names to differentiate their products and promote exports. Fruit exports have a particularly promising future. Fruits of a great variety and quality have long been a staple of Taiwan, in both the Japanese and ROC eras, and more than 30 varieties of fruit now account for one-fourth of the island's total agricultural production. Although fruit imports into Taiwan have increased, the fruit industry is working hard to promote its products.[49]

One highly promising new approach to farming in Taiwan is organic farming, which has already begun to make significant inroads in the US and other countries but is still in its infancy in Taiwan.[50] Organic farming, properly practised, prohibits the use of synthetic pesticides, chemical fertilizers, pharmaceuticals, preservatives, genetically modified organisms, growth stimulants, and irradiation. Practitioners of organic farming also must be certified by government approved organizations authorized to carry out the tests necessary to grant certification to a farmer. Organic products are not necessarily more nutritious, but they are guaranteed to be free of any chemical residues and to be all natural and environmentally friendly. Because of the much higher labour input required, especially in a subtropical environment like Taiwan's with all of its insects and plant diseases, and the fact that production is small-scale, organic products inevitably cost more in retail markets. Nonetheless, just as in the US, Taiwan's increasingly affluent people, especially the well-educated and upper-middle class urbanites, are beginning to develop a taste for organic products, and farmers in Taiwan are beginning to respond to what is definitely still a niche market, but could become a large-scale market in coming decades.[51]

Initial experiments with organic farming in Taiwan date back to 1987 (an interesting correlation with the beginning of Taiwan's political renaissance), but commercial production for markets did not get under way until about 1995. Formal organic standards were first issued by the government in 1999, although livestock, poultry and dairy products are not yet covered. Hence, organic farming is limited so far to crops. As of 2003, the total area of farmland certified for organic farming was just over 1,000 ha, a tiny fraction of the island's total 850,000 ha of farmland. Rice accounts for over half of all organic produce, which also includes leafy vegetables and tubers, fruits, tea, sugar cane and yams. All are for domestic consumption, although there is hope that Taiwan's farmers might eventually get into exporting their organic produce. That is probably quite a few years down the road though, at best.

As is true in the US and elsewhere, many so-called 'organic' products in the markets are not truly organic, and that is a problem in Taiwan as well. There are five government-assisted certification entities in Taiwan, three of them accredited by the Council of Agriculture. The government is also working on organic standards for livestock and poultry, which are more complex products to certify than crops. Vegetarians remain a large component of organic produce consumers, and the challenge for producers and marketers of such products is how to convince the general public that it is in their best interest to go organic. True believers hope that eventually all farming on the island could be converted to organic, but that is probably unrealistic.

Nonetheless, Hualien County is not going to wait for the rest of Taiwan to catch up. Agriculture plays a much bigger role in the economy in Hualien, which has very little industry, as already noted earlier in this volume. Hence, Taiwan's entry into WTO has hit Hualien and its farmers harder than some other parts of Taiwan. Thus, in 2004 the Hualien County government established a core policy to create a zero-pollution environment in the county by promoting organic farming. A number of pioneers in organic farming have proven that the higher costs involved are offset by higher incomes. So the government hopes that organic farming might be the answer to Hualien's agricultural woes.

In the broader environmental realm, the government had the following goals in place as of 2005, for the purpose of developing 'eco-agriculture' and promoting 'sustainability of resources'.[52]

1 Reduce the numbers of farms and promote recycling to raise pollution prevention efficiency. Promote correct fertilization and elevation farmland effectiveness. Encourage no-fishing days with incentives, to protect offshore fishery resources. Improve irrigation on farmland to maximize agricultural water use efficiency.
2 Enhance reforestation and sustainability of forests and maintain biodiversity. Promote full-scale mountain management, hazard prevention and environmental beautification, and accelerate post-disaster rehabilitation and reforestation to safeguard national land. Reconstruct botanical gardens to maximize their multiple functions for research, education, and leisure.
3 Improve the general development of agriculture and upgrade the life of farmers and fishers.
4 Promote life-long learning for farmers and build quality agricultural manpower resources. Improve rural life quality and set up production support systems to mobilize rural communities.
5 Guide fishermen groups toward sustainable management and enhanced services. Improve the agricultural financing system, install agricultural banking outlets nationwide and maintain agricultural financing order.

6 Conduct agricultural and fishery welfare and aid, and enhance public-guided administrative services to upgrade agricultural policies.

These broad goals appear to contain more rhetoric than substance. No one can really argue against such seemingly noble objectives. The devil is in the details, though, as the expression goes. Exactly how to implement these goals is the question, although the new National Land Plan discussed above, and still being debated, does address some of these goals with specific policies and regulations. The next step after enactment, though, is enforcement, and this is where Taiwan historically has been weakest, as we have already noted in many parts of this book. In many ways, Taiwan is at a watershed in its land use policies and practices. The future of the island's quality of life, perhaps even the system's survivability, is at stake.

Ongoing environmental issues facing agriculture

In addition to the complex land use issues facing both slopelands and lowlands, as discussed above, agriculture in general faces a variety of ongoing environmental issues that will not be soon resolved, if ever. These have been covered already to varying degrees in other parts of this book. The following is a wrap-up of each of them, as they currently stand.

Land subsidence

As already noted in Chapter 2, land subsidence is a persistent and serious problem in various parts of Taiwan. The problem is caused by over-pumping of groundwater, at the rate of 400 million cubic metres a year for the whole island. That is about twice the storage capacity of Shihmen Reservoir, Taiwan's third largest, in Taoyuan County. The most serious excess pumping of groundwater occurs in Yunlin, Changhua, and Chiayi Counties in central west Taiwan (the driest area of the island). The water is used for irrigation, for fish ponds, and by industry.[53] Although over-pumping has been reduced over the past ten years (down from a peak of 1.5–2 billion cubic metres in 1996), the government still needs to do more. For example, total depth of land subsidence in Yunlin reached 2.24 m in 2004 and continues at a rate of 10.6 cm per year, second only to the 14.2 cm per year in Changhua. Altogether, about 40 per cent of Yunlin County is continually sinking, an astonishing amount.[54] To slow down or even halt the problem, the government is embarked on a programme of closing public wells in the region, which are believed to cause 70–80 per cent of the problem. However, water consumers in the area will need to find alternative sources, or change professions. The government hopes to entice farmers raising fish in ponds to do just that, by finding other means of earning a livelihood. The many industrial estates in the region are not going to be shut down (although more efficient use of water is certainly being promoted), so agriculture is likely to carry the brunt of the sacrifice (as it tends to do worldwide in any competition between agriculture and urban/industrial consumption). The government has another urgent motive in stopping the land subsidence: the new high speed rail line runs right through this region, and its operability is being threatened by the continuing subsidence.

Water shortages

Partly as a result of the land subsidence problem noted above, water conservancy is a major thrust of government efforts in Taiwan today, especially in the south-west coastal plain, which suffers from chronic water shortages. That region receives the lowest rainfall annually in Taiwan, but is a prime agricultural area as well as important industrial region, as already discussed earlier in this book. As we noted in Chapter 3, government efforts to build new reservoirs in the region have met with mixed reactions from environmentalists, who argue that, no matter how desperate the water needs in the south-west (or elsewhere, for that matter), the government needs to carefully examine the total water balance of the entire island before embarking on construction of ecologically questionable new reservoirs.

One approach to the problem has been the Council of Agriculture's programme, 'Methods and Procedures for Ecological Engineering and Management in Farm Irrigation'.[55] The first stage, carried out in 2002–4, consisted of studying more efficient methods for farmland water conservancy and testing those methods on the construction of dams, water paths for irrigation and drainage, reservoirs, and ponds. The second stage, to run 2004–7, consists of making sure the methods tested in the first stage really work and local people understand and trust the methods. The third stage, implemented in 2006, involves long-term analyses and checks, to monitor the various water conservancy measures, for the purpose of eventually establishing a sustainable system that will use water to maximum possible efficiency. Hence, it will be a number of years before we will know the outcome of this programme, but it holds promise for the future well-being of Taiwan's agricultural system.

Agricultural water usage already has decreased in Taiwan. In 1997–8, water usage for the whole island came to 18 billion cubic metres (m^3), of which agriculture consumed a whopping 70 per cent, or 12.2 billion m^3 (farmland usage was 10.6 billion m^3, breeding 1.5 billion m^3, and animal husbandry 116 million m^3). The COA projects that total agricultural water usage will decline further to 10.6 billion m^3 by 2011.

The controversy surrounding the proposed Hushan Reservoir project, first proposed back in 1979, illustrates the difficulties in resolving this thorny issue of water shortages.[56] Located in Yunlin County, the project has been touted by the Water Resources Agency (Ministry of Economic Affairs) as designed to solve the water shortages of residents of Changhua, Yunlin, and Chiayi counties, and to reduce the rate of land subsidence. Opponents, including members of the Cabinet's National Council for Sustainable Development, have argued that the reservoir is actually designed to supply water to the naphtha cracker projects and steel companies in the area, is located in a geologically unsafe site, and would be silted up within 20–50 years. Moreover, the critics question why the project has not had to undergo a thorough environmental impact assessment. Environmentalists versus developers; the struggle goes on.

Deteriorating water quality in reservoirs

One part of the water shortage problem noted above is making sure that the water available in Taiwan is of high enough quality to meet the island's needs. The ongoing problem of deteriorating water quality in some of Taiwan's reservoirs was briefly noted in Chapter 3 also. The problem mainly is with older reservoirs, such as Shihmen in Taoyuan County. Altogether, the TEPA claims there are now six reservoirs (out of 57 total) still in trouble. The problem arises mainly from the fact that these reservoirs were built during a period of much more lax rules about use of land in reservoir watersheds. Hence, pig raising produces wastes, construction and farming contribute to soil erosion, and pollution from chemicals flows into the water system. The government has made progress on curbing these activities, as we have noted, but there is still unfinished business.

Pig raising and water pollution

Pig raising has definitely been one of the major agricultural contributors to water pollution in Taiwan over the decades, as noted in earlier chapters. The government is fully aware of the problem, but has faced something of a dilemma in that pigs are a vital part of Taiwan's rural economy, a major source of income to many farmers and agribusinesses, and pork is a major source of protein in the average Taiwanese diet. Hence, it is unrealistic to expect Taiwan to abandon raising of pigs. Instead, the government has tried to adopt a compromise approach, through reducing the numbers of pigs, particularly in environmentally sensitive areas such as watersheds of reservoirs, and to encourage farmers with smaller operations to phase out pig raising and leave the industry mainly to larger-scale producers who (theoretically) have the economies of scale to employ the soundest environmental practices in disposal of waste products. Moreover, with Taiwan now in the WTO, it makes sense to focus on larger-scale pig raisers who can stay competitive with imported pork products. This definitely is an environmental problem that can be controlled and virtually eliminated with proper management and enforcement of regulations.

Toxic chemicals in soil and water

Also in Chapters 2 and 3 we examined the more insidious problem of toxic wastes buried in various sites that result in often extremely serious health hazards to humans and livestock. Most of these dumpings took place in the martial law era when environmental regulations were lax or non-existent, and the dangers just became apparent in recent years. Unfortunately, however, the practice still continues to this very day to some extent by unscrupulous businesses. The area of farmland officially noted by the TEPA as polluted by toxic chemicals seems relatively small (a few thousand hectares) in relation to the roughly 850,000 ha of total farmland on the island. Nonetheless, these toxic waste sites have wreaked havoc on farm communities and small towns in many parts of the island, especially the west coast plain where most of Taiwan's people live and industry is concentrated. Hence, the area of cultivated land lost to cultivation (at least temporarily) by toxic wastes gives a misleading impression of the true dimensions of the problem. This truly is a ticking time bomb for Taiwan, for all citizens, rural or urban.

Fisheries problems

Taiwan's fisheries problems were examined in some detail in Chapter 3 also. This is a troubled industry due to a variety of factors, including: (1) pollution of inland and coastal waters that affects fisheries output and marketability of products; (2) intense international competition and growing regulation to save deep-sea fish stocks from depletion, while Taiwan has steadily expanded to become one of the world's major deep-sea fishing participants, sometimes amidst controversy; (3) changing diets and lifestyles among Taiwan's increasingly urbanized population, leaving behind a way of life more attuned to an era long ago and that appeals to fewer and fewer people.

Conclusion

The issue of land use, particularly in rural areas, remains a critical issue for Taiwan and affects everyone on the island. Land use issues trouble every state around the world, of course, but they take on added intensity in those places with high population density, such as Taiwan. Hence, trying to determine what is the 'best' use of each type of land is a thorny question that has consumed planners, officials, and citizens of Taiwan for many decades, and likely will continue to do so indefinitely into the future. There will never be unanimous agreement on what constitutes the best use of land, because the interests of different groups in Taiwan's complex society are too varied. The best that can be hoped for is compromise agreement on usage of land that offers the greatest benefits to society as a whole. Unfortunately, in Taiwan's highly charged political environment, even achieving a compromise consensus becomes a formidable challenge.

5 Reflections

Toward a green silicon island

Taiwan celebrated the 20th anniversary of its 'rebirth', so to speak, in 2007, marking the date when martial law was lifted and the democratization and flowering of human rights began to take place. The year of 1987 was also momentous on the environmental front, as we have seen, with the formal establishment of the central government's Environmental Protection Administration, followed by a complex hierarchy of agencies and organizations at all levels, whose collective goal has been to work toward a cleaner and better environment, a higher quality of life, for the people of Taiwan. While some of the foundations for this campaign had been laid before 1987, that year truly was a watershed on the environmental front, in that it opened the floodgates to policies, programmes, and initiatives on a broad front, public and private, from which the people of Taiwan are beginning to reap the benefits. Much progress has been made, without question. However, it has been uneven progress, and certainly not as rapid as many people in Taiwan wished for. Bad habits and attitudes toward the environment are difficult to change. Moreover, it must be remembered that Taiwan labours under a multitude of constraints or burdens, as outlined in previous chapters (especially Chapter 1), including: (1) limited land area (especially lowlands); (2) very high population density and hence also very high density in numbers of factories, automobiles, and all the other consequences of high levels of urbanization and industrialization; (3) a subtropical, typhoon-prone climate that exacerbates the liabilities of steep slopes and an unstable geologic structure; (4) an immature political democracy made more dysfunctional by complex and politicized ethnography and the inability of many people to focus on the future rather than the past; and (5) an implacable nemesis across the Strait in the Peoples Republic of China that does everything possible to make Taiwan's political separateness uncomfortable. Given all these burdens, it is a wonder that Taiwan has accomplished as much as it has on the environmental front.[1]

The environment: successes and failures

What then have been the greatest successes? Opinions on this undoubtedly would vary, depending on whom one asks. People have different viewpoints as to what is most important in the environment and what contributes the most to quality of life. For example, officials at the TEPA might give different answers compared with those from ordinary citizens polled on the street. In the opinion of the authors of this book, the greatest successes so far have been the following.

Nature conservation

Because of its geographic location and characteristics, Taiwan has some of the most spectacular natural wonders to be found in Asia. Taiwan has set aside fully one-fifth of all its land to establish a system of national parks, and other kinds of conservation areas, to protect the island's unique land, water, fauna and flora for future generations to enjoy. These preserved areas not only enhance the quality of life of Taiwan's residents, but provide an invaluable resource for Taiwan's efforts to promote tourism. This is a remarkable record, and a system of nature conservation that other nations could well emulate.

Reducing air pollution

The air over Taiwan is cleaner today, in city and countryside, than it was 20 years ago. Taipei now has among the cleanest air to be found in Asia's major metropolitan centres. Even the southern city of Kaohsiung, with its formerly notoriously foul air, has seen big improvements. These major strides have been accomplished through a variety of effective measures including: cleaner gasoline; use of catalytic converters; getting rid of the most polluting forms of motorcycles; strict monitoring of motor vehicle emissions; improvements in public transportation, including mass rapid transit systems (as in Taipei) that are helping to wean people away from reliance on private vehicles (cars and motorcycles); better traffic law enforcement; much stricter enforcement of emissions standards for factories, construction, and businesses.

Disposal of solid waste (garbage)

There has been impressive progress in changing the entire process of handling garbage and changing peoples' attitudes and behaviour toward garbage disposal. In some respects, Taiwan is one of the leaders in the world in handling this troublesome and messy byproduct of modern civilization. This is especially evident to anyone who is familiar with the old days in Taiwan, with the smelly piles of rotting garbage along the streets. Now Taiwan has a relatively efficient system of pickup, recycling, incineration, and hygienic disposal of garbage.[2] Plastic bags and Styrofoam are becoming a plague of the past. Taiwan is a healthier, more attractive place because of the improvements.

Conversely, Taiwan has not been as successful in a number of other areas of environmental protection, in spite of sometimes well-meaning efforts and some progress. The most serious deficiencies still are as follows:

Improving land use

In spite of decades of attempts to legislate controls and improve land use practices, major problems persist. This is especially true in the slopelands, but also in the coastal zone and river basins, as Chapter 4 detailed. As is typical of most societies around the world, it took natural disasters in recent years (unusually severe typhoons and earthquakes) to cause enough destruction, loss of life, and disruption of the economy to finally convince both public and government that procrastination and ineffective enforcement of laws and regulations had to come to an end (the parallel with Hurricane Katrina in 2005 and its impact in the US is striking). It seems to be a characteristic of societies around the world that they do not act forcefully enough until confronted with a major crisis. Taiwan may be on the verge of a new era in its land use policies and practices, in both slopelands and lowlands. But environmentalists cannot breathe easy yet.

Meeting Taiwan's energy needs

Taiwan has a huge and growing appetite for energy, much like that of all highly industrialized, urbanized societies around the world. Taiwan has relied, up to now, primarily on coal, natural gas, and oil as fuels for thermal power plants and motor vehicles, as well as for industry, supplemented by some hydropower, a little bit of exotic or renewable power (such as wind), and a modest experimentation with nuclear power. All of these energy sources have their drawbacks. There is no cost-free energy source. But for Taiwan the costs, versus the benefits, seem to be greatest for coal, oil, and natural gas, because of the cost of importing the raw materials, the volatility of world supplies, and the air pollution and residues left over after burning. Nuclear power has alluring benefits, but a huge potential cost in the risks of nuclear contamination from accidents and/or waste products. Renewable energy sources are unlikely to account for much for the foreseeable future, even though the government has nobly dedicated itself to their usage (especially wind power). Hence, the people of Taiwan, like those in every country around the world, must face the harsh reality of their energy needs and reach common agreement on how much they are willing to compromise to find the right energy balance for their circumstances. None of the solutions will be ideal.

Controlling GHG emissions

Even though Taiwan is a second-string player in the global economy, it produces a disproportionate share of greenhouse gases (GHG), especially CO_2. Being outside the 'family' of nations (through the United Nations) makes it all the harder, though, for Taiwan to participate in international efforts to control GHG and global warming. It is to Taiwan's credit, nonetheless, that the government has committed itself to playing a constructive role in such efforts. To achieve success, though, requires improvement in a broad range of other environmental efforts, to find the right mix of energy sources, to improve efficiency in factories, motor vehicle emissions, and other areas.

Meeting water quality needs

Gaining control of the problem of sewage disposal, and improving the quality of Taiwan's potable water supply still remain unattained goals. Too many rivers and streams are still polluted. Progress has been made, such as cleaning up the Kaoping River basin in southern Taiwan and its pig raising, but it has been slow and very uneven. Tap water anywhere in Taiwan still really is not safely drinkable directly from the tap (notwithstanding claims to the contrary from some government officials).[3] Far too high a proportion of sewage still flows into rivers and the sea completely untreated. For complex political and financial reasons, it is proving very difficult and slow to get piped sewage systems, and sewage treatment plants, built in all of Taiwan's cities and towns, where the need is greatest. This task will take decades to complete, if it ever is. The problems of the Shihmen Reservoir are illustrative of the links between land use and water quality, and the difficulties in changing them.

Meeting water quantity needs

The quantity of water available in Taiwan also is still a largely unresolved issue.[4] In spite of relatively high annual rainfall, Taiwan's water provision is out of balance, as the result of: (1) uneven distribution of available water over the island; (2) extreme fluctuations in the availability of water during the year (droughts alternating with deluges of rain from typhoons and other storms); (3) flooding and destruction caused by steep slopes and rapid run-off; and (4) absurdly low prices for water supplied to urban and rural residents, thus encouraging people to waste water. There is nothing the government can do about when precipitation will fall, but there is a great deal more that it can do in channelling and using that water more effectively. This is a very long-range endeavour also.

Disposal of toxic wastes

This is a buried time bomb, with innumerable sites all over the island, some of them still unknown, hiding all kinds of toxic residues of modern industrial society (not even including nuclear waste here). The TEPA has been trying to tackle these, but they are among the most difficult to accurately determine and clean up. Moreover, the task puts the TEPA in direct conflict with some private businesses, some of which are extremely adept at skirting laws and regulations and turning a blind eye to social responsibilities. The real long-range solution to the problem is in educating the business community in the virtues of running their enterprises in an eco-friendly way. This will take a long time. Progress is slow. In the mean time, environmentalists hope Taiwan will never have its Minamata.[5]

Toward a green silicon island?

Will Taiwan ever become the 'Green Silicon Island' that the government aspires to? It is such an illusive goal, in the face of daunting problems and challenges, and without even agreement about what exactly it means. The Democratic Progressive Party chose the colour green as its symbol because green represents purity of spirit and incorruptibility, as well as respect for environmental quality, the foundation stones of the party and everything it stood for. Thus, when the DPP finally came to power with the Presidency of Chen Shui-bian in 2000, it adopted the slogan 'Green Silicon Island' as a catchy name to symbolize what the party stood for and to win over voters. The idea, vaguely expressed, was to somehow restore and protect Taiwan's environment while still maintaining the island's powerful industrial economy, focusing on high tech industry (i.e. the 'silicon' chip), which had become the mainstay of the economy by the end of the twentieth century. A noble goal, one that most industrial nations around the world also aspire to, even if they do not use this exact slogan (although many governments now use the word 'green' in their development plan slogans). Indeed, one would be hard pressed to find a government anywhere that does not wish it possible to promote economic growth (whether centred around information technology or any other sector) and still somehow miraculously not despoil the environment. The two goals almost seem inherently contradictory and doomed to conflict with each other. It is the nature of the modern world we live in. And yet, in Taiwan as in many other countries, there are political parties, organizations, and people who absolutely believe in the feasibility of combining economic growth with environmental protection in a form of *modus vivendi*. It is called sustainable development.

Sustainable development

The concept of sustainable development first emerged in the early 1980s and spread around the world. Many nations now ostensibly promote this strategy, or at least give lip service to it, and many have governmental units dedicated to promoting it. There are numerous definitions of the term, but perhaps one of the simplest is that adopted by the United Nations World Commission on Environment and Development, which defined sustainable development as: 'Development that meets the needs of the present without compromising the ability of future generations to meet their own needs.'[6] That definition is a big enough umbrella to accommodate just about any interpretation imaginable. More specifically, however, sustainable development can be interpreted as the intersection, or overlapping, of the three main sectors of environment, economy, and society of which all nations or states are composed, as portrayed in the top half of Figure 5.1. In other words, a sustainable state (or nation) is one in which the three sectors must be healthy and cannot degrade any one of the other two sectors. The ultimate goal is for the three spheres of environment, economy, and society to be completely overlapped and fully integrated with each other.[7]

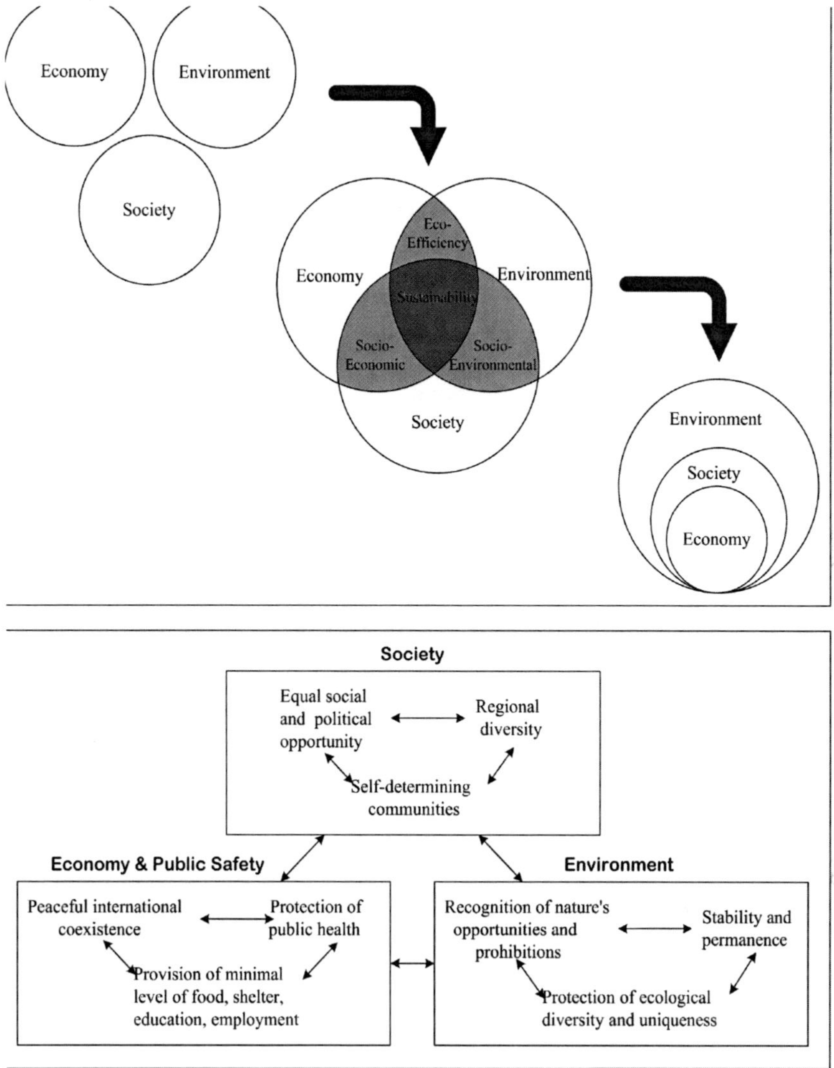

Figure 5.1 Sustainable development (source: Authors).

The UN, in its 1987 report, *Our Common Future*, further suggested five key concepts that distil the essence of sustainable development:[8]

1 The needs of the future must not be sacrificed to the demands of the present.
2 Humanity's economic future is linked to the integrity of natural systems.
3 The present world system is not sustainable because it is not meeting the needs of many, especially the poor.

4 Protecting the environment is impossible unless we improve the economic prospects of the earth's poorest peoples.
5 We must act to preserve as many options as possible for future generations, since they have the right to determine their own needs for themselves.

The link between sustainable development and environment thus is very clear. However, environmental protection by itself is not a complete definition of sustainable development. Rather, sustainable development is a holistic concept, with hints of utopianism in it, suggesting that human society can do a much better job than it has so far in organizing and developing the world in which we live.

Another interpretation of sustainable development used by some environmentalists in Taiwan is shown in the bottom half of Figure 5.1. This is basically a variant form of the concept in the top half, in that it suggests the need for a holistic integration of a society's economic, political, and social development with environmental protection, and for all societies (nations/states) to do so in harmony with each other.[9]

Taiwan's approach to sustainable development, expressed through the 'Green Silicon Island' slogan, derives from growing dissatisfaction with what is seen as an ecologically damaging, low-cost, high-waste, export-led pattern of development begun during the KMT era and continued, in the eyes of critics, in the DPP era (in spite of protestations by the government to the contrary).[10] Specifically, a number of very troubling factors are seen by critics as red flags or indicators of failure to make adequate progress toward the green silicon island: per capita emissions of carbon dioxide; the quality of reservoirs and the effective amount of water resources in reservoirs; reduction of arable land; reduction in the share of natural coastline; per capita production and use of cement; rising use of pesticides and other agricultural chemicals; the ratio of water-intensive manufacturing operations compared to total manufacturing output; the share of environmental-related spending in the central government budget; the number of automobiles in urban areas; the decline of green areas.[11] The foregoing list is by no means inclusive either.

As we noted in Chapter 1, Taiwan's economy has gone through a momentous evolution over the decades. As one study put it, the decades could be characterized as: canned food in the 1950s, textiles in the 1960s, petrochemicals in the 1970s, information technology in the 1980s, and semiconductors in the 1990s. By the early years of the new century, Taiwan ranked among the top three producers worldwide for more than 30 high tech products, with almost 30 per cent share of the world's integrated circuit production value, just slightly behind that of the US. Nonetheless, Taiwan is seen as relying too much, up to now, on original equipment manufacturing (OEM) and original design manufacturing (ODM), rather than producing brand name products with higher returns for Taiwan, in both earnings and image.[12] To put it another way, Taiwan's economic growth averaged 8.23 per cent during the 1950s, 9.15 per cent in the 1960s, 10.23 per cent in the 1970s, 8.15 per cent in the 1980s, and just 6.37 per cent in the 1990s.[13] An admirable record, to be sure, but also showing clear signs of a maturing and slowing economy, and a need for change. By 1989, Taiwan's 'heavy, chemical, and technology-intensive industries' (chemical raw materials and products, rubber and plastic products, basic metals, metal products, machinery, electronics, electric appliances, transportation vehicles, and precision instruments) made up 65 per cent of the manufacturing industry's total production value. Yet, the industrial sector as a whole was contributing only 30 per cent to Taiwan's GDP by 2001, versus 67 per cent from services.[14] Taiwan's recession at the start of the new century was seen as the result of several key factors: (1) the slowdown in the global economy; (2) the economy's overdependence on the so-called high-tech industries; (3) the continued migration of Taiwan's manufacturers to China; (4) the problem of bad loans in the financial sector; and (5) the domestic political gridlock.[15]

To be fair, the development path for an economy as large as Taiwan's is not easily re-routed. Change takes time. Moreover, the Chen Shui-bian government has been severely undercut by the Pan-Blue opposition and its obstructionist behaviour. The political gridlock has all but paralyzed the government. Yet, at the same time, the central government and the DPP too often have chosen political expediency over principles, and have dismayed many by seeming incompetency in many areas of policy making and enforcement.[16] To compound matters, cases of alleged corruption have surfaced among various figures in the DPP and even in Chen Shui-bian's family. Corruption is seen as an endemic problem in both political camps, causing Taiwan's ranking in international corruption indices to slide steadily downward. Whether true or not, these scandals tend to diminish trust in the government by the public. There were even calls in 2006 for Chen's impeachment or that he should resign. The noble principles that Chen and the DPP campaigned on seemed to have been abandoned.[17] Small wonder that many former DPP supporters and environmentalists lost faith. An air of frustration and worry has gripped much of Taiwan's society in recent years and that is likely to affect future decision-making in environmental management, regardless of the outcome of the 2008 presidential election.[18]

Finding the right plan

The government's perspective on this was initially reflected in the latest Six-Year National Development Plan, known as Challenge 2008. This was launched in 2002, not long after Chen came into office and after holding an Economic Development Advisory Conference in 2001 that brought together the leaders of Taiwan's economy to help the new government chart the future path for Taiwan. Challenge 2008 was supposed to serve as guidance to the public and private sectors in Taiwan, and to help Taiwan transform itself into a green silicon island.[19] The plan was expected to cost some US$75 billion and to have the following specific goals: (1) expanding to 15 the number of products and technologies that meet the world's highest standards; (2) doubling the number of foreign visitors; (3) increasing R&D expenditures to 3 per cent of GDP; (4)

reducing the average unemployment rate to less than 4 per cent; (5) increasing the average economic growth rate of the next six years to over 5 per cent; (6) increasing the number of broadband Internet users to over six million; and (7) creating approximately 700,000 jobs. These goals, heavily focused on economic growth and job creation, help explain the public criticisms of the government's interpretation of 'sustainability'. By 2007 none of these goals had been achieved.

The plan had 11 major areas of emphasis that covered a broad spectrum of topics and objectives in what critics felt was actually a hodgepodge and poorly thought-out effort to get Taiwan out of recession and back on track. The areas of emphasis included: (1) increasing the English capability of Taiwan's young people, to help further globalize the island; (2) developing the cultural creativity industry to further promote the knowledge-based economy; (3) developing peripheral industries to enhance Taiwan's position in the Chinese-speaking world; (4) developing an international base for R & D to bring Taiwan up to the level of the most advanced nations; (5) increasing value-added production in a broad range of high-tech industries; (6) doubling the number of international tourists visiting Taiwan (currently a little over one million a year, the largest number coming from Japan); (7) digitizing Taiwan, much the way that Singapore and Hong Kong have done; (8) developing Taiwan as an operations headquarters for Taiwan businesses as well as multi-national corporations; (9) further improving the transportation infrastructure; (10) conserving water resources and the ecology (the only explicitly environmental-focused objective); (11) constructing new home-town communities, to revive small towns (including aboriginal villages).

In the four years since Challenge 2008 was launched, however, the economy has not done particularly well and support for the six-year plan weakened both within and without the government. Although the plan remains nominally in place as guidance for Taiwan, in reality both the public and private sector have been searching for a better solution.

Some hoped that solution would come out of a new national planning conference held in Taipei, 21–22 April 2006, known as the National Conference on Environmental Action Towards Sustainability. This event was proposed and organized by the National Council for Sustainable Development under the Executive Yuan (see Chapter 3), and with participation by relevant government agencies, and was attended by some 3,000 participants.[20] The TEPA was responsible for handling the logistics. The conference focused on a variety of themes: major transportation projects; development of water resources; reduction in greenhouse gas emissions; industrial restructuring; nuclear power.[21] Local environmental groups, such as the Taiwan Academy of Ecology, the Green Party of Taiwan, and the Taiwan Watch Institute, protested the focus of the conference, contending that it was primarily directed at economic sustainability, not environmental protection, and held public demonstrations during the conference.[22] Organizers of the conference, however, reminded everyone that the agenda had been drafted from the bottom up through numerous meetings and forums of environmentalists, local residents, academics, and officials of local government agencies over a six-month period.[23] Inevitably, on a topic as broad and controversial as sustainable development, it would be impossible to satisfy everyone.[24]

The conference participants, nonetheless, managed eventually to reach consensus on 267 specific items, but failed to do so on another 53 items.[25] Among the most controversial topics were the future of nuclear power, and setting an exact timetable for reducing carbon dioxide emissions to comply with the Kyoto Protocol.

On the nuclear power issue, Lee Yuan-tseh, head of the prestigious Academia Sinica, Nobel Laureate in chemistry, and probably the most respected intellectual figure in Taiwan, spoke at the conference and frankly told the group that Taiwan really has no choice but to embrace nuclear power for the immediate future, probably for another 50 years or so, until such time as alternative energy sources can be expected to come on line. Lee also argued that continued use of nuclear energy would also make it easier for Taiwan to meet clean air standards and CO_2 emissions reductions, by lessening Taiwan's dependence on fossil fuels for energy. Taiwan's emission of 12.4 tons per person annually of CO_2 is exceeded only by the US and Australia, he claimed. Thus, Lee further recommended that the Fourth Nuclear Plant had to go forward. Critics countered that Taiwan still has not resolved the vexing problem of nuclear waste management, and the safety of Taiwan's nuclear plants has yet to be verified.[26]

The Fourth Nuclear Power Plant remains a lightning rod for the anti-nuclear movement. As noted in previous chapters, this controversial project has been on and off for the past 20 years, and a political football in the nuclear power issue. In spite of huge cost over-runs and countless delays for the project, the Chen government in 2006 reiterated its reluctant support for completing this project, even though the DPP has publicly re-stated that it remains committed to a nuclear-free Taiwan in the long run. Assuming the plant's construction goes ahead without further interruption, the No. 1 generator of the plant is scheduled to go into operation by July 2009, the No. 2 generator one year later, and the entire project to be completed by 2012.[27] As for the nuclear waste issue, that still remains unresolved. The legislature did pass legislation requiring the government to get the approval of people living in areas where nuclear waste storage sites are planned. About 97,000 barrels of waste are currently stored on Lanyu Island, but a new, safe site must be found within five years, according to the new legislation.[28] No matter what site eventually is chosen, critics of nuclear power probably will never be satisfied, as they maintain there is no such thing as a 'safe' site for nuclear waste. Like many countries around the world, thus, Taiwan is going to continue to wrestle with the thorny issue of energy for many decades to come. Somehow, the right balance between the various forms of energy that works best for Taiwan's circumstances has to be found. That balance has so far eluded everyone.

The issue of CO_2 emissions and Taiwan's participation in reducing global warming is obviously tied in with the energy issue discussed above, as well as the future restructuring of Taiwan's economy. Here, also, there are deep divisions of opinion within Taiwan's society as to the future development path Taiwan ought to take.

This was painfully evident in yet another national planning meeting, the Conference on Sustaining Taiwan's Economic Development, held in Taipei 27–28 July 2006, scarcely three months after the previous conference. This one, though, was clearly focused primarily on economic development, as opposed to the April conference that had a strong environmental tilt. Originally suggested by President Chen in 2005 as more than just a follow-up to his 2001 conference (which was geared mainly to the short term, to overcoming the recession following the 1997–8 Asian fiscal crisis), he wanted this July conference to be both short-term and long-term in its focus and to lead to real results that could shake Taiwan out of its doldrums. Instead, the conference seemed to accomplish little except to reinforce special interest groups and opposing political factions in their fiercely protected camps. Political gridlock seemed destined to rule the day yet again.

The conference this time was attended by only about 200 key business leaders and government figures, with a sprinkling of academics and NGO leaders, who soundly criticized the conference even before it started.[29] A number of environmental protection groups, such as the Taiwan Environmental Protection Union, even dropped out of the conference on its first day, complaining that the meeting was actually 'anti-sustainable'.[30] The conference, led by Premier Su Tseng-chang, basically ended up being gridlocked by distrust from industries and a boycott from the Taiwan Solidarity Union and environmental groups.[31] Consensus was reportedly reached on over 500 issues, but critics countered that the really important matters were effectively tabled because they proved too controversial, namely, relations with China. Specifically, the conference failed to reach agreement on: (1) establishing direct transportation links with China; (2) lifting the ceiling on China-bound investments (currently capped at 40 per cent of a listed company's net worth); (3) allowing banks to set up branches and subsidiaries across the Strait and investing in Chinese financial institutions; (4) the trading of the Chinese currency in Taiwan; and (5) allowing Chinese banking companies to set up representative offices in Taiwan.[32] From an environmental viewpoint, one of the most curious things to come out of the conference was what appeared to be possibly a new slogan under consideration by the government, in that the participants did reach consensus on building what was labelled 'a Green Value-Added Island', again trying to link environmental protection with economic growth, but not limiting growth to just high-tech industry. Whether this new slogan will come to supplant 'Green Silicon Island' remains to be seen; the original is far catchier than the newer version. Regardless, slogans alone are not going to produce the results needed.

The conference did spell out specific statistical goals for various facets of Taiwan that are worth noting. From a base in 2005 of 4.09 per cent economic growth, per capita GDP of US$15,271, and a jobless rate of 4.13 per cent, the goal was set at an average of 5 per cent growth per year, per capita GDP of US$30,000 and an unemployment rate of below 4 per cent by 2015. In addition, industrial restructuring would boost the service sector to 76.5 per cent of GDP by 2015 (from 73.56 per cent in 2005). Yet, participants failed to agree on a target volume for reduction of GHG or any measures to achieve that goal.[33]

Taiwan's ageing population may be a big factor in whether or not those goals are achieved. As of 2006, almost 10 per cent of the population was over the age of 65; that will rise to 13 per cent by 2015, and 20 per cent by 2025. The birth rate fell to 1.1 children in 2005, one of the lowest rates in the world, and the dependency ratio is now 7.2 to 1. Thus, one of the agreements in the conference was on the need to establish a national pension plan law. There is also a growing gap between rich and poor that needs addressing.[34] These social and demographic trends do not augur well for the economic goals laid out.

Perhaps the most scathing indictment of the conference was voiced by former President Lee Teng-hui, now a strong supporter of the TSU, strongly anti-China and anti-business, and increasingly critical of Chen Shui-bian. Lee said that he found it 'scary' that people who did not understand economics were determining the country's economic course, and that the conference had been 'kidnapped' by the business sector, eager to expand economic relations with China.[35] Whether one agrees with Lee's strong opinions or not, he did focus attention on the reality of Taiwan's current situation, i.e., a fierce struggle between those who see increased ties with China as the key to Taiwan's future, and those who feel quite the opposite.[36] Regardless of growing economic linkages with China,[37] the reality over recent years has been for a growing propensity of people in Taiwan to view themselves as 'Taiwanese' rather than 'Chinese', to regard Taiwan as a fully sovereign and independent state, and to strongly support the idea that Taiwan's future should be decided by the people of Taiwan and no one else.[38]

Whatever development plan Taiwan finally comes up with, the science-based parks will be a critical part. The first one, the Hsinchu Science-Based Industrial Park, just south of Taipei, was established at the end of 1980. The Southern Taiwan Science Park was founded in 1996 in Tainan County. Planning for the third major park, the Central Taiwan Science Park in Taichung, started in 2004. These three parks are regarded as essential to the goal of achieving a green silicon island.[39] The science and technology push is also evident in the government's ten-year Environmental Science and Technology Parks (ESTPs) Project. This calls for the creation of four areas to have so-called 'eco-industrial parks'. These will be environment-friendly industrial parks that are supposed to have minimum or no secondary pollution. Two of these parks are already under planning and construction. One is in Kaohsiung County, in a complex that will serve the petrochemical, metal, and electronic sectors. The second is in Hualien County and oriented toward a biotechnology and life sciences centre with focus on renewable resources and energy, and bio-agricultural industry. Two other future parks will be in Taoyuan and Tainan Counties.[40]

Counterpoints

Yet, at the same time that these high-tech efforts are being vigorously pushed, some of the foundations of the old economy continue to be promoted, creating tensions and political infighting. As of 2005–6, one of the biggest controversies swirled around proposals for the petrochemical and steel industries from the Ministry of Economic Affairs, which got the green light from the Executive Yuan, on the rationale that these industries are essential national industries that supply raw materials for downstream players, and that Taiwan was not yet ready to phase out those sectors and had no choice but to proceed with the projects.[41]

The Kuokuang Project

The Kuokuang Project, also known as the Yunlin Petrochemical Technology Park, is one of the key economic projects being supported by the government in recent years. Reportedly one of the largest petrochemical investment projects in Asia, the US$1.3 billion complex is a joint venture of Taiwan's eight petrochemical companies and Taiwan's state-run China Petroleum Corp., which holds a 43 per cent stake. The complex will occupy more than 2,000 ha (20 km^2) on the Yunlin coast and comprise oil refineries, petrochemical plants, cogeneration plants, and an industrial harbour. When fully operational in 2015, the complex is projected by its promoters to generate an annual production value of US$11 billion, create more than 25,000 jobs, and account for almost one per cent of Taiwan's total GDP. In order to proceed with the project, the developers will need to buy some 800 ha (8 km^2) of privately owned land on the site.[42] No small undertaking, with potentially huge benefits to the economy. Hence, the government's support is understandable. Nonetheless, passing the EIA in the next few years will not be easy for this project.

FPG steel mill

Right next door to the Kuokuang Project is the site for a new steel mill, proposed by the Formosa Plastics Group (FPG), which, as we have noted before, has figured prominently in a number of controversial environmental issues over the years. FPG claims this new steel mill would produce 7.5 million metric tons of steel annually and exert a strong multiplier effect on the economy, leading to an additional production of US$12 billion worth of goods and services, directly and indirectly create some 40,000 jobs, and attract around US$3 billion worth of investments in local auto parts manufacturers and other metal-production sectors.[43]

Critics, including some in the government, have their doubts, though, about the real need for such a steel mill. Taiwan's state-run China Steel Corp. produces about 16 million tons of steel a year, and private companies turn out another two million or so tons. Taiwan also imports more than seven million tons per year. However, existing mills are expanding their capacity, and demand in Taiwan is expected to dwindle, not increase, as more and more manufacturing shifts offshore (especially to the PRC). Moreover, Taiwan can easily afford to import steel to make up any shortfall.

Environmentalists dislike both projects and have argued that the government was promoting an out-of-date industrial policy that would increase CO_2 emissions by 23 million tons per year, or nearly 10 per cent of the island's total CO_2 emissions in 2000.[44] This has given pause to the TEPA, which also has heightened concerns about the source of water to be used by the new steel plant, the amount of emissions to be produced, and the impact of waste gases and other toxic substances.

In other words, these two mammoth projects symbolize the current dilemma facing Taiwan's government, regardless of which political party is in power, and the inherent contradictions that exist in trying to mesh the goal of a green silicon island with the immediate need for jobs, raw materials, and economic growth. No matter which way the government, whether under Chen or his successor, swings on these two highly controversial projects, some groups will be infuriated. Taiwan's business community frequently speaks up to remind the government, and environmentalists, of the dangers of going too far on the environmental front with what they regard as unrealistic, unfair, or unenforceable policies that risk driving businesses, including foreign investors, out of Taiwan.[45] The government, under whatever party, has no choice but to pay careful attention to such warnings.[46]

Taiwan's future

The World Economic Forum (WEF) publishes an annual *Global Competitiveness Report*, which is based on eight factors: openness of an economy; role of the government; development of financial markets; quality and quantity of infrastructure; quality of technology; quality of business management; efficiency and flexibility of labour markets; and the quality of judicial and political institutions. These factors are all directly or indirectly related to an economy's political environment. The sum of these eight factors is a Global Competitiveness Index (GCI). The WEF's GCI is widely regarded as one of the most reliable rankings of the world's nations in how they are performing as national economies. International corporations and investors pay attention to the WEF report in making their business decisions. Hence, it matters to national governments where their nation stands in the rankings, because it provides a barometer of how competitive their nation is in the global economy.

In 1999 Taiwan was ranked 4th globally, after Singapore, the US, and Hong Kong, and was 3rd in Asia. In the 2004 WEF report, Taiwan ranked No. 4 among 104 countries and came out on top in Asia, retaining the No. 1 position in competitiveness in the region (Table 5.1). By 2004, Singapore and Hong Kong had fallen to No. 7 and No. 21, respectively (the PRC dropped from 32nd to 46th in the same five years).[47] In the 2005 WEF ranking, Taiwan slipped one place to 5th in the GCI, while Japan fell to 12th place, South Korea rose from 29th to 17th, Singapore rose to 6th, Hong Kong slipped further to 28th place, and China did even more dismally, falling from 46th to 49th.[48] Hence, from the perspective of the WEF ranking, Taiwan looks quite good.

From another perspective, however, Taiwan has reason to be concerned about

Table 5.1 World economic forum growth competitiveness index, 2004–5

Country	*GCI 2005 rank*	*GCI 2005 score*	*GCI 2004 rank*
Finland	1	5.94	1
United States	2	5.81	2
Sweden	3	5.65	3
Denmark	4	5.65	5
Taiwan	5	5.58	4
Singapore	6	5.48	7
Japan	12	5.18	9
South Korea	17	5.07	29
Hong Kong	28	4.83	21
China	49	4.07	46

Source: www.weforum.org/site.

its economic future. Over the past ten years, there has been a steady erosion of Taiwan's economic position in East Asia, especially in comparison with South Korea. By 2005, per capita income in Taiwan was about US$15,600, while South Korea's reached almost $16,300, the second year in a row in which South Korea topped Taiwan, reversing a dominance Taiwan had enjoyed for many decades. Taiwan's per capita income is now the lowest of Asia's 'Four Little Dragons' (South Korea, Taiwan, Hong Kong, Singapore). Since 1992, Taiwan's per capita income has increased an average of only about US$270 per year.[49]

Within Asia, thus, South Korea shows the most dynamic improvement in the region, further demonstrating how it has managed to catch up with Taiwan in many ways and overcome the setbacks following the 1997 Asian fiscal crisis. Nonetheless, in spite of much soul searching among leaders and intellectuals in Taiwan, the island is still in a relatively strong position in Asia. But the clock is ticking. The island's relative strength is slowly slipping away, as the WEF report and other data suggest. To reverse that trend, the island and its people need to come together. Unfortunately, the current political climate does not encourage optimism on that possibility.[50] But without unity, Taiwan will have great difficulty charting the path that will blend economic growth, environmental protection, and social progress together in the right combination, in order to achieve a 'Green Silicon Island'.

Hanging over all of this, of course, is the grave danger from China, with its implacable determination to incorporate Taiwan into the PRC, at whatever cost proves necessary. All the nuances and complexities of Taiwan's internal economic, political, and social systems become moot issues in the face of that larger threat. Trying to predict the potential impact of reunification on Taiwan, including on its environment, is another realm of investigation beyond the scope of this book, especially if that reunification is achieved with military force. We can only hope that calmer heads prevail in this struggle.

The *2005 Sustainable Development Index* issued by the NCSD provides a sobering review and assessment of where Taiwan stands at the present time.

First issued in 2003, the Index serves as a means of observing progress while providing cautionary advice and direction for policy makers. In the Index, a total of 40 core indicators are categorized under six dimensions, as follows:[51]

1 Environmental pollution: With 1988 as an index of 100, overall environmental quality has not noticeably improved, although individual components of that environmental quality have made progress, such as waste resource recycling rates.
2 Ecological resources: Composite index scores (again, with 1988 = 100) have been on a continual downward slide since 1996, reflecting the ever-increasing intensity of population and pressure on resources.
3 Social pressure: This index has shown a trend toward sustainability, due to such factors as increased recycling rates, improved unemployment rates, and others.
4 Economic pressure: This index has also shown a trend toward sustainability, which is encouraging.
5 Institutional response: There has been steady progress since the early 1990s in this dimension, indicating the obvious fact that the government has been giving ever greater priority to environmental issues and moving Taiwan toward sustainability. In other words, the will and effort are there; the results, however, do not always turn out as well as hoped for.
6 Urban sustainable development: The long-term trend in this index is also positive, indicating movement toward sustainability, based on such factors as urban per capita income, use of mass transport, rate of urban expansion, etc.

The ultimate value of an index like this, of course, is the degree to which the results are used by government and public to make improvements. Indices are nothing more than yardsticks. Taiwan has come this far in improving its environmental quality and developing a sustainable system, but there is still a long road ahead.

Notes

1 A crowded island: Taiwan's environmental setting

1 Parts of this chapter are based on themes introduced by Williams in his 2003 study of the Taiwanese diaspora (Williams, 2003a).
2 Throughout this volume, the term 'state', rather than 'country' or 'nation', is used in reference to Taiwan, because of the political limbo in which the island continues to exist. Strictly speaking (at least in the eyes of a majority of the world's governments), Taiwan is not recognized as an independent nation state, even though it functions quite effectively as one. This complex political issue is discussed in more detail later in the volume.
3 Unless otherwise noted, the statistical data for this chapter are taken primarily from the *Taiwan Yearbook*, published annually by the Taiwan government. The other main source is Hsieh's detailed geography, which remains the basic geographic text about Taiwan, especially for physical and historical geography (Hsieh, 1964).
4 Technically, this land area figure includes the other islands administered by the government of Taiwan (or the 'Republic of China', as it is still formally known): the Peng-hu Islands (Pescadores) off the south-west coast; the islands of Matsu and Kinmen, just off the coast of Fujian province; and a few other small islands off the east and south-east coasts, including most importantly Lanyu Tao (Orchid Island), Lu Tao (Green Island), and some others. Collectively, however, these other islands account for a small fraction of 'Taiwan's' total area.
5 To convert millimetres to inches, multiply by 0.03937. Annual precipitation for Taiwan thus varies from as little as 40 inches on the central west coast to as much as 180 inches in the rainy north-east.
6 Strong earthquakes occur by the dozens each year, with countless lesser quakes. In 2001, for example, a total of over 16,200 quakes were detected, of which 136 were large. The biggest quake that year had a magnitude of 6.8 on the Richter scale. The four largest quakes in Taiwan's (recorded) history took place in 1848 in Changhua, in 1906 at Meishan in Yunlin/Chiayi, in 1935 in Hsinchu/Taichung, and in 1999 in Nantou ('The historical impact of earthquake damage on Taiwan society and culture', *Taipei Journal*, 26 October 2001, p. 7.).
7 See, for example: Bain (1993); Gold (1986); Ho (1978); Hsiao *et al.* (1989); Hsiung *et al.* (1981); Kuo *et al.* (1981); Ranis (1992).
8 Liu (1987).
9 Currently, this term is often romanized as *Holo*, rather than the older *Hoklo*.
10 The term 'Mainlander' also technically refers to other mainland Han immigrants who have come to Taiwan in more recent years, such as spouses arriving in the post-1987 period of liberalization and improved relations with the PRC.
11 Known as *Kuo Yu (Guo Yu)* (or 'national language') on Taiwan, and *Putonghua* (or 'common speech') in the PRC. The problem of romanizing Chinese words is histori-

cally a vexing problem, and still not totally resolved. The mainland uses the *Hanyu pinyin* romanization. Taiwan, for political reasons, initially stuck to the old Wade-Giles system (from the colonial era) but in recent years has been experimenting with other forms, so that today, unfortunately, one encounters a hodgepodge of romanization systems on the island. Romanization of people's names has become especially irrational. Right now, there is a movement to promote *Tongyong pinyin*, which is a modification of the *Hanyu pinyin* used in the PRC. Tongyong pinyin spelling is being gradually phased in throughout Taiwan, but in the meantime street signs and other places where proper names are shown can be very confusing to foreigners. The central government made the mistake in 2002 of officially adopting Tongyong, but then left it up to city and county governments to decide whether to use it or not. The result is confusion, for locals and foreigners alike. (See: 'Editorial: drowning in alphabet soup', *Taipei Times*, 25 October 2004, p. 8). Because of the confusion and incomplete use of Tongyong pinyin on Taiwan, for this volume we are sticking as much as possible with traditional Wade Giles romanization (albeit simplified, without hyphens and apostrophes). However, for some names we are forced to use the current spelling. The result is that the reader will, regrettably, sometimes see various romanization forms used in this volume, in both text and on the maps. Our hands are tied.

12 The official, politically correct term today is 'indigenous' people, but for simplicity sake, and to differentiate the aborigines from the Taiwanese, also indigenous to the island (in a historically relative sense), the term 'aborigine' is used throughout this book. ('Editorial: what's in an aboriginal name?', *Taipei Times*, 27 January 2006). Figure 1.6 was produced before the 13th and newest aboriginal group (the Sakiraya) declared itself independent.

13 For the most complete and authoritative analysis of all aspects of Taiwan's demography, see Selya (2004).

14 'Scientists question wisdom of boosting birthrate', *Taiwan Journal*, 28 July 2006, p. 2.

15 'Taiwan heading for aging milestone, must face challenges beginning now', *Taiwan Journal*, 24 November 2006.

16 'Ministry of Interior proposes subsidies as birth rate declines', *Taiwan News*, 6 June 2004.

17 See, for example: Ahern and Gates (1981); Allee (1994); Bing (1986); Chen *et al.* (1991); Knapp (1980); Meskill (1979); Rubinstein (1999); Shepherd (1993).

18 A key source of first-hand information and fascinating historical detail of the late Qing/early Japanese era is Davidson (1903).

19 Camphor is a volatile, crystalline substance with a strong odour, derived from the wood of the camphor tree, and used to protect fabrics from moths, in manufacturing cellulose plastics (including early photographic film), and in medicine.

20 One of the foremost authorities on the Japanese era and the beginning of the ROC era was George Kerr, whose two books, *Formosa Betrayed* (1965) and *Formosa* (1974) are required reading for understanding those critical years in Taiwan history.

21 A narrow gauge railway linking Taitung and Hualien was completed by the Japanese, but rail connections between the east side and west side of the island had to wait until the 1980s and beyond.

22 This parallels a similar change in public attitude that has occurred in the US in recent decades, with many people now questioning the wisdom of the vast dam and water resource exploitation programmes in the American West in the twentieth century (see Reisner, 1986).

23 For a brief history of Ali Shan's role in Taiwan's history, and the building of the famous railway by the Japanese and what happened to it, see D. Momphard, 'Taiwan's alpine rail', *Taiwan Review*, June 2005, pp. 16–19. The Japanese were particularly interested in the tree that produced *hinoki*, a prized wood used by the Japanese for building baths, temples, and imperial palaces. Some of these trees were

1,000 years old or more. Mount Ali still is a prime recreation/tourism site, and now has a modern road to the top that has greatly expanded the volume of visitors, putting greater pressure on the area's resources.

24 Hsieh (1964), pp. 168–70.
25 Williams (1980).
26 Much has been written about the tumultuous and tragic years of the period, 1945–9. In addition to Kerr (1965), a basic reference is Lai, *et al.* (1991).
27 Clough (1978); Liu (1987).
28 Calkins *et al.* (1992). The '2–28 Incident', as it is called, began as an effort by native Taiwanese to seek redress of grievances against the KMT forces and ended in a bloodbath by the KMT forces to wipe out all organized opposition to KMT domination of Taiwan. The aftermath was to poison relations between Mainlanders and Taiwanese for decades to come. The bitter feelings still linger, and may never entirely disappear.
29 Shaw and Williams (1991).
30 The term NIE was especially pushed by China and its supporters, who argued that Hong Kong and Taiwan were not 'countries', or independent states, and hence the politically neutral term, 'economy', was more appropriate to describe them in this context.
31 Actual investment by Taiwan in the mainland has been estimated as high as US$100 billion since it began in the 1980s, much of it unrecorded and technically not approved by the Taiwan government.
32 *Statistical Yearbook of the Republic of China 2005* (Taipei: 2006). Taiwan's economic miracle seemed to be waning by the end of the 1990s, as factories relocated to the mainland and unemployment rose in the wake of the Asian financial crisis of 1997. By 2001, Taiwan's economy was actually contracting for the first time in half a century. By 2004, however, the corner was turned and an economic upswing was under way.
33 See, for example: Corcuff (2002); Kerr (1974); Lai *et al.* (1991); Lee (1999); Tien (1989); Wachman (1994).
34 The likelihood of the latter objective diminished in practical terms as the years went by. Certainly by the post-Mao era (the 1980s onward), as China became ever stronger, only the most die-hard KMT members still had any lingering hopes of recovering the mainland. Recovery of the mainland, and the fiction that the ROC government was the sole legitimate government of all of China, were finally abandoned publicly in the early 1990s.
35 Two books that recount the legacy of that era are: Simon and Kau (1992) and Rubinstein (1994). Bello and Rosenfeld (1990) take a broader look at the darker side of development not only in Taiwan but also South Korea and Singapore.
36 To be fair, by the early 1980s the government was coming around to the realization that the environment mattered and the impact of development projects on the environment needed to be considered. Hence, the concept of Environmental Impact Assessments (EIA) began to be debated in the Executive Yuan and other forums, as the decade of environmentalism took shape (see Chapter 3).
37 Williams (2003b).
38 Williams (1981).
39 Williams *et al.* (1983).
40 TEPA *Yearbook 2004*.
41 Yen Chia-kan, a political lightweight with no political ambitions, served as Interim President during the transition period, 1975–8, between the two Chiangs, father and son.
42 Acquiring the niceties of democracy takes some time, however. The Taiwan legislature has often been the scene of physical violence and hot tempers as members of opposing parties contended over policies and control of the government. Governmen-

tal efficiency sometimes suffered as a result, with grid deadlock paralyzing the political process in a polarized legislature. Few if any Taiwanese, however, wished to return to authoritarian rule; most simply hoped that politicians could learn to live with each other and get things done for the public welfare, not for their private interests. This remains a key issue today, including in the environmental arena.

43 Williams (1994).

44 SEF and ARATS have not met since the late 1990s, however, because of PRC antipathy toward the Chen Shui-bian government.

45 The organization is formally known in Taiwan as simply the 'Environmental Protection Administration' (not Agency), but for the sake of clarity, to avoid confusion with the EPA in the US, throughout this volume we are using the acronym TEPA (Taiwan EPA). Officials in the ROC government sometimes used to refer to the organization as the ROCEPA, but this is rarely encountered outside of political circles. Given the controversy regarding the future status of the use of 'ROC' in reference to Taiwan, this acronym (ROCEPA) is not used in this study.

46 Chen (1994); Williams and Chang (1989; 1994 reprint).

2 The perils of development: Taiwan's environmental problems and their causes

1 This chapter is based, in part, on previous studies by Williams and Chang (1989) and Williams (1992, 1996).

2 'Public dissatisfied', *Taipei Times*, 12 February 2006.

3 R. Dobson, 'Environmental pluses and minuses', *Taiwan Business Topics* (Taipei American Chamber of Commerce), 27 November 2002, p. 30.

4 V. Chang, 'Quantifying quality', *Free China Review*, February 1999, pp. 30–7; Williams (1996).

5 One actually might get some disagreement among the public in Taiwan right now about quality of life issues, with some arguing that the sharp political divisions between the Pan Blue and Pan Green coalitions and their supporters, which has brought legislative progress to a near halt, detracts more from the quality of life than even environmental problems (see Chapter 3).

6 'Taiwan among worst in environmental quality', *China Post*, 24 February 2003.

7 'Government urged to focus on sustainability', *Taipei Times*, 1 February 2006; 'A fair report on the environment', *Taipei Times*, 12 February 2006.

8 National Council for Sustainable Development (2006), p. 40.

9 There is a huge literature in the environmental realm. As examples of representative publications issued just in 2004 and relevant to various facets of this study, see: Bigio and Dahiya; Booth; De Blij *et al.*; Dyson *et al.*; Economy; Gordon *et al.*; Harris; Jasanoff and Martello; Myers and Kent; Perez; Randolph; Sims; Smil; Sorensen *et al.*; Speth; Sutton; Trefil; Wisner and Blaikie.

10 Huddle *et al.* (1975).

11 He (1991); Smil (1984, 1993, 2004).

12 In some parts of Asia, such as Hong Kong, the Air Pollution Index (API) is used, but it has essentially the same standards as the PSI.

13 By contrast, however, API readings in the high 100s have been recorded in the past few years in parts of Hong Kong, the result of heavy local traffic and pollutants drifting over from factories in Guangdong. ('Hong Kong air pollution hits a record high', *China Post*, 10 September 2004.)

14 TEPA *Yearbook 2004*: 'Air quality improvement'.

15 Selya (2004), p. 226.

16 See I. Underwood, 'Environment: airborne menace', *Free China Review*, September 1996. Selya (2004) devotes a whole chapter in his book on Taiwan's demography to the themes of 'mortality, morbidity, and public health'.

17 TEPA *Environmental Yearbook* 2004, 'Population, vehicles, and factories', Table 7–1.
18 Underwood (1996), pp. 28–36.
19 TEPA briefing 2001, 'An experiment on intensive monitoring of China's dust storms'; plus unpublished data from the TEPA.
20 'Warning issued on major dust storm in Taiwan', *Taiwan News*, 20 April 2005.
21 'Strong sandstorm from China to affect Taiwan today, says EPA', *Taiwan News*, 20 March 2006.
22 A. Batson, 'China's choke-hold over Asia', *Far Eastern Economic Review*, 8 July 2004; D. Murphy, 'Stinking, filthy, rich', *Far Eastern Economic Review*, 21 October 2004.
23 TEPA website, 2004, 'Table 3–1, 'Pollution level of rivers'.
24 W. Chou, 'Kaohsiung City's Love River brought back from the dead', *Taipei Journal*, 23 August 2002, p. 4.
25 TEPA website, 2004, Table 4–1 'Collection and disposal of municipal solid waste'.
26 P. Li, 'Environment: trash transfigurations', *Taipei Review*, October 2000, pp. 46–53.
27 TEPA website, 2004, Table 4–3, 'Properties of municipal solid waste'.
28 Ironically, the US is increasingly moving toward the use of plastic bags, and away from paper bags, at the very time that Taiwan (and many other countries) is doing its best to do away with plastic bags.
29 C. Fanchiang, 'Taiwan EPA reports on success of plastics ban', *Taiwan Journal*, 11 July 2003, p. 2.
30 TEPA website, 2004, Table 7–3, 'Livestock production'.
31 TEPA website, 2001, 'Background'.
32 TEPA website, 2004, Table 3–6, 'Wastewater pollution reduction'.
33 TEPA website, 2001, 'Current status of industrial waste handling'.
34 TEPA *Yearbook 2001*: 'Industrial waste management'.
35 'Whose backyard is big enough? Seeking solutions for hazardous industrial waste', *Sinorama*, July 1999, pp. 114–27.
36 'Waste being dumped all over Kaohsiung County, BEP says', *China Post*, 3 September 2000; 'EPA promulgates stricter rules on the disposal of toxic chemical waste', *China Post*, 22 September 2000.
37 'Spotlight thrown on waste handling', *Taipei Times*, 13 February 2006.
38 'Politicians accused of dumping toxic soil on riverbanks', *China Post*, 16 October 2001.
39 'EPA to investigate toxic rice paddies', *Taiwan News*, 17 October 2001.
40 Y. T. Chiu, 'Villagers sick of pollution, literally', *Taipei Times*, 12 October 2004, p. 2.
41 E. Hoh, 'Environment: facing up to a dirty secret', *Far Eastern Economic Review*, 12 December 2002; 'EPA clarifies pollution zone around RCA's Taoyuan plant', *Taiwan News*, 6 July 2005.
42 'EPA confirms cases of alarming pollutant levels', *Taiwan News*, 17 August 2005.
43 A dramatic and tragic demonstration of the effects of dioxin on humans was provided in 2004 during the campaign for President of Ukraine, when the opposition candidate, Victor Yushchenko, was found to have been deliberately poisoned with dioxin, leaving him severely scarred and ill, with likely long-term impact on his health.
44 'Tainan factory site blamed for cancer', *Taipei Times*, 28 October 2004; 'EPA head visits polluted industrial site in Tainan', *Taipei Times*, 13 July 2005; 'Residents protest dioxin pollution by jumping into river', *Taiwan News*, 24 August 2005.
45 'Tainan City seeks funds from government in dioxin case', *China Post*, 13 July 2005.
46 'Secret pipes pollute river in the south', *Taipei Times*, 17 March 2005.

47 'EPA says animal feed may be source of toxin in eggs', *Taiwan News*, 28 September 2005; 'COA insists dioxin scare now under control', *Taiwan News*, 29 September 2005; 'Taiwan Steel found responsible for dioxin pollution', *Taiwan News*, 17 December 2005; 'Duck farmers slam EPA's handling of pollution case', *Taiwan News*, 20 December 2005.

48 'Fish contains 47 per cent persistent organic pollutants', *China Post*, 23 December 2000.

49 'EPA finds alarming levels of mercury in some fish', *Taiwan News*, 21 December 2005; 'Fishermen protest warning by EPA against eating fish', *Taiwan News*, 24 December 2005.

50 'Taipei residents mercury content is too high, tests say', *Taipei Times*, 30 August 2005.

51 'EPA report sees drop in mercury levels', *Taiwan News*, 9 January 2006.

52 M. Keliher, 'Environment: exporting a problem', *Far Eastern Economic Review*, 12 December 2002.

53 'Whose backyard', op. cit.

54 In fact, however, China is actively seeking trash of all kinds from many countries in the world, and has become the biggest importer of other peoples' trash, especially plastics, paper, board, and other waste. With its cheap labour, China recycles the trash and uses the components as raw materials for its manufacturers or for export, causing recyclers in other countries to complain of being driven out of business. Most of this recycling is done in small plants with minimal or non-existent environmental standards, especially for keeping track of toxic materials. ('Imported garbage threatens China', *Guardian*, London, 21 September 2004) Also see: Economy (2004).

55 'Malay company in toxic waste scandal could face charges', *Taipei Times*, 10 June 2004; 'Malaysia returns waste', *Taipei Times*, 9 June 2004.

56 M. Ivins, 'Texas woman shows true grit', *Lansing State Journal*, 29 October 2005.

57 'EPA urged to toughen battery recycling', *Taiwan News*, 2 October 2003.

58 'EPA to tighten monitoring over used auto batteries', *China Post*, 17 September 2003.

59 'Battery dump underlines larger recycling problem', *Taipei Times*, 15 February 2006.

60 As just one example, Bangalore, the centre of India's IT economy, reported that in 2005 that city would generate a total e-waste of 1,000 tons of plastic, 1,000 tons of iron, 300 tons of lead, 0.23 tons of mercury, 43 tons of nickel, and 350 tons of copper, most of which would be improperly disposed of in illegal dump pits for e-waste, which are not currently covered by Indian environmental laws ('Bangalore set to choke under tons of e-waste', *Taiwan News*, 1 November 2004). Sadly, some countries, such as the US, try to dispose of obsolete or unwanted computers and peripherals by shipping them to other countries, including India, as 'donations', which are legal under the *Basel Convention On The Transboundary Movements of Hazardous Wastes and Their Disposal*, which came into force in 1992 and was ratified by 159 countries (but not by the US) ('Toxic e-waste threatens India's technology center', *Taipei Times*, 1 November 2004.)

61 'International e-waste generation', www.ewaste.ch/facts_and_figures (accessed on 9 June 2005).

62 B. Jhu, 'EU green laws trigger change in electric, electronics industries', *Taipei Journal*, 15 April 2005, p. 3.

63 'China only the first to move on RoHS in Asia', www.ewaste.ch/services/news/ (accessed on 6 June 2005).

64 TEPA website, 2004, 'Noise control in Taiwan'.

65 This group preference of Chinese society is reflected also in a much greater

tolerance for crowding and high density of living in urban areas, and the cities reveal this in the design and placement of buildings and other facilities.

66 TEPA website, 2004, 'Draft amendments to noise act'.

67 Williams (2003b).

68 P. Courtenay, 'Taiwan's cities hustle to boost quality of life', *Taiwan Journal*, 18 June 2004, p. 7.

69 C. Fanchiang, 'Exhibition touts government land-use policies', *Taiwan Journal*, 3 December 2004, p. 2.

70 'Asia's best cities: quality of life index for Asia's key cities', *Asiaweek Magazine*, 15 December 2000. This weekly magazine was published in Hong Kong, but is now defunct.

71 By comparison, Shanghai and Beijing ranked 15th and 16th, respectively, in the overall ranking.

72 The May, 2007 issue of *Taiwan Review* has a special focus on the built environment, and presents an interesting overview, with many illustrations, of what is good and what is bad about Taiwan's built environment.

73 See, for example: 'Ecology: damage control', *Taiwan Review*, March 2006, pp. 40–5.

74 Strictly speaking, Taiwan also creates some new land through land reclamation along the coast for new industrial sites. However, Taiwan's use of this method of ameliorating the shortage of level land pales in comparison to the efforts of states such as Japan, Hong Kong, and Singapore, which are world leaders in land reclamation.

75 *Agricultural Statistics Yearbook*, 2004, pp. 178–9.

76 *Taiwan Yearbook* 2003, 'Agriculture'.

77 *Agricultural Statistics Yearbook* 2004, p. 179.

78 *Taiwan Yearbook* 2003, 'Agriculture'.

79 F. Li, 'Farmers face challenges of WTO membership', *Taipei Journal*, 11 January 2002, p. 8; W. Y. Chiu, 'Taiwan agriculture facing difficult restructuring', *Taipei Journal*, 17 May 2002, p. 8; P. Courtenay, 'Free trade in agriculture leads to concern over domestic food supply', *Taipei Journal*, 16 August 2002; F. Li, 'Plight of island's agricultural sector addressed at national conference', *Taipei Journal*, 13 December 2002, p. 3.

80 P. Gao, 'Water/soil conservation: stopping the slides', *Taipei Review*, 4–11 June 2002.

81 Native Taiwanese and aborigines are much more likely to chew betel nuts than are Mainlanders or their descendants, or better educated, higher income people of all ethnic groups.

82 Gao (2002), op. cit.; 'Betel nut chewing proves a tough habit to crack', *Taiwan News*, 6 April 2006.

83 *Agricultural Statistics Yearbook* 2004, p. 86.

84 A Lee, 'Betel-nut farming area reaps what it sows', *Taiwan Journal*, 3 September 2003.

85 A survey in 2004 found that one out of every four working people in Taiwan smoke, and nearly half of all workplaces do not enforce a no-smoking regulation. The survey found that about 42 per cent of men, but only 3 per cent of women, smoked regularly ('One out of every four professionals in Taiwan smoke, survey suggests', *Taiwan News*, 6 December 2004). Still, these are much lower figures than found in the PRC.

86 'Taipei County hopeful Wang makes issue over environment', *China Post*, 12 November 2001.

87 Gao (2002), op. cit.

88 O. Chung, 'Water/soil conservation: taming a river', *Taipei Review*, 12–17 June 2002.

89 'Asian typhoons take unusual turn in worst season on record', *China Post*, 13 September 2004. The year 2004 was one of the worst on record for typhoons in the Northwest Pacific. A total of 19 typhoons were recorded (the all-time record was 24 in 1971), several of which hit Taiwan, China, Korea, and Japan with great destructive force. Interestingly, the east coast of the US, especially Florida, was also hit by an unusual number of hurricanes in the same season.
90 O. Chung (2002), op. cit.
91 Davidson (1903), pp. 558–9.
92 W. Chou, 'Once and future resource planted on Arbor Day', *Taipei Journal*, 12 April 2001, p. 4.
93 Hsieh (1964), pp. 298–9.
94 Forestry Bureau webpage (www.forest.gov.tw/web/English2).
95 Hsieh (1964), pp. 301–2.
96 Forestry Bureau webpage, op. cit.
97 'Snake Alley' in the old Wanhua district of Taipei has long been a popular site for residents and visitors alike, where one can see live snakes (some poisonous) taken from cages, strung up on a hook, and skinned alive. The blood is then drained off, the gall bladder removed for medicine, and the blood drunk from a glass by males seeking an aphrodisiac, the act performed with great machismo in front of spectators. The snake meat is then cooked in various dishes. This macabre sideshow is favoured mainly by less well-educated, very traditional Taiwanese males, but in some ways it reflects the attitude of too many people in Taiwan toward the island's fauna. Snake Alley has been dressed up in recent years for the tourist trade, but a few of the snake shops continue to thrive, in spite of public disdain by environmentalists and others.
98 'Conservationists decry mistreatment of captured birds', *Taiwan News*, 3 November 2004; 'Freeing captured birds criticized', *Taipei Times*, 3 November 2004.
99 'Illegal hunting rife in the Dawushan nature reserve', *Taipei Times*, 31 March 2006.
100 'Poaching leads to new Tan-ta Forest Trail rules', *Taipei Times*, 10 July 2006, p. 1.
101 J. Sneddon, 'Good news for Taiwan's endangered species', www.sinica.edu.tw/tit/environment (2001).
102 'Farmers up in arms over damage caused by protected monkey', *Taipei Times*, October 7, 2004.
103 'Infected monkeys raise alarm in Kaohsiung City', *Taiwan News*, 6 June 2004.
104 Numerous such problems have occurred with other species elsewhere in the world, such as the reintroduction of wolves to Yellowstone National Park, which has been controversial with ranchers.
105 'Prevention center for red fire ants to be launched today', *Taiwan News*, 1 November 2004.
106 'Center geared to stamp out red fire ants in three years', *Taipei Times*, 2 November 2004.
107 'COA to tackle foreign weed species', *Taipei Times*, 14 October 2004.
108 Hsieh (1964), pp. 291–5.
109 'Nation needs a fisheries ministry', *Taipei Times*, 6 December 2005.
110 'Taiwan's tuna-fishing industry gets flak', *Taipei Times*, 18 June 2004.
111 'Japan admits to overfishing bluefin tuna by 25 per cent', *Taipei Times*, 3 March 2006.
112 'Fins with the blues', *The Economist*, 27 January 2007.
113 'Yuhsun No. 2 sets sail to help guard against overfishing', *Taipei Times*, 12 August 2006, p. 2.
114 'ICCAT restores Taiwan's tuna quota for 2007', *Taiwan Journal*, 8 December 2006.
115 TEPA *Environmental Yearbook* 2004, p. 279.
116 *Taiwan Agricultural Yearbook*, 2004, p. 200.

117 *Time* magazine had a special issue on the topic on 3 April 2006 that made for disquieting but informative reading.
118 A. Regalado and J. Ball, 'As planet heats up, scientists plot innovative fixes', *Wall Street Journal*, 22 October 2004.
119 J. Helprin, 'Bush stands fast in rejecting Kyoto treaty', *Lansing State Journal*, 7 November 2004.
120 Regalado and Ball, op. cit.
121 A. Regalado, 'When a plant emerges from melting glacier, is it global warming?', *Wall Street Journal*, 22 October 2004.
122 TEPA webpage, 2004, 'Response to climate change in Taiwan, ROC'.
123 'More effort urged on curbing emissions', *Taipei Times*, 3 November 2004.
124 TEPA website, 2001, 'Ozone layer protection in Taiwan'.
125 TEPA website, 2004, 'Ozone layer protection: what Taiwan is doing'.
126 TEPA website, 2001, 'Current air pollution problems'.
127 M. Lu, 'Controversy over nuclear plant still raging', *Taipei Journal*, 22 September 2000.
128 'Taiwan can't afford a Chernobyl disaster, experts warn', *China Post*, 31 August 2000.
129 M. Lu, 'Fate of 4th nuclear plant up in the air', *Taipei Journal*, 19 May 2002, p. 1.
130 M. Lu, 'Opposition reacts strongly to plant decision', *Taipei Journal*, 3 November 2000, p. 1.
131 M. Pao, 'Taiwan: nuclear fallout', *Far Eastern Economic Review*, 25 January 2001, p. 29; M. Lu, 'Project restart eases tension but no-nuke message clear', *Taipei Journal*, 23 February 2001.
132 R. Y. Lin, 'No-nukes forum mulls green energy sources', *Taiwan Journal*, 4 July 2003, p. 1.
133 The Matsu group of islands is a holdover from the still technically unfinished Chinese Civil War, in which the ROC forces managed to maintain control of the Chinmen and Matsu Islands off the coast of Fujian in 1949, and maintained military bastions there throughout the decades. In the rapprochement era since the late 1980s, the PRC has not pushed for recovery of those islands, and Taiwan is now in the process of turning them into tourist sites and conduits for trade and other linkages between Taiwan and China. The islands remain a curious anachronism from a troubled past (see Chapter 1).
134 C. Fanchiang, 'New timetables okayed for construction of facility to replace Orchid Island nuclear waste depository', *Taiwan Journal*, 20 February 2004.

3 Back from the brink: resolving Taiwan's environmental problems

1 Various studies show that while environmentalism is beginning to awaken in the PRC, as NGOs begin to emerge and the seeds of activism are being planted, the political system still makes it extremely difficult, if not outright dangerous, to openly confront the government. See, for example: Cooper (2003), Ho (2001), Lawrence (2003).
2 One of the most important studies of that momentous decade of the 1980s was the *Taiwan 2000* report, a massive first-time effort by a group of concerned scholars and scientists to put together a comprehensive analysis of Taiwan's environmental situation, and to propose solutions to at least some of the problems. Published in English, as a private publication by the committee involved, the study had a big impact and eventually came out in a Chinese language version (Hsiao *et al.*, 1993). Moreover, the government TEPA adopted the study as a basic reading requirement for its employees. Although much has changed since its release in 1989, the report remains an essential reference work for understanding Taiwan's environment. The

contributors are a 'who's who' of leading scholars and writers in the environmental field in Taiwan and elsewhere, including Chang Ch'ang-yi, Chiang Pen-chi, Chu Yun-peng, Michael Hsiao, and Lucia Severinghaus, who constituted the Steering Committee.

3 As one example: Linda Gail Arrigo (former wife of Shih Ming-teh, a successful DPP candidate elected in 1992 to represent the city of Tainan in the National Legislative Yuan and later in the mid-2000s an outspoken critic of Chen Shui-bian), has been a strong critic of the government's environmental record, and wrote a scathing critique of the situation in Tainan City that remains vivid reading today (Arrigo, 1994). Also see: Arrigo *et al.* (1996).

4 S. R. Severinghaus, 'The emergence of an environmental consciousness in Taiwan', unpublished paper presented at 1989 Annual Meeting of the Association for Asian Studies, Washington, DC; Hsiao (1999).

5 Hsiao (1999); also see: Hsiao (1989); Hsiao and Tseng (1999).

6 Kim (2000); Tang and Tang (1999).

7 The leader of the so-called rebellion, Shih Hsing-men, a chemical engineer at National Taiwan University, went on to lead in the establishment of an NGO, the Taiwan Environmental Protection Union, in 1988 as a result of the rebellion.

8 Reardon-Anderson (1992).

9 Jinn-yuh Hsu, a geographer at National Taiwan University, devoted an entire paper to this famous incident of environmental activism in the 1980s (Hsu, 1995).

10 The incidents at Three Mile Island and Chernobyl in 1985–6 certainly did not help Taipower's case either.

11 Some of the legal actions are still pending or in appeal, but in most cases the lawsuits have been settled in favour of the plaintiffs, i.e. the people affected by the contamination. The buildings affected included schools, apartment buildings, and office buildings. The Taiwan Atomic Energy Council (TAEC), whose responsibility is to monitor and control these kinds of problems, was initially very passive and unresponsive during the KMT era, but eventually started to take stronger action, and the government eventually arranged physical examinations and compensation for victims. However, the contaminated structures have not been torn down or rebuilt, and hence the contamination remained an unresolved problem as of 2005. See: Kao (2002).

12 'Activists seek to halt building of nuclear site', *Taiwan News*, 5 January 2005. Actually, the KMT itself, even before Chen was elected President in 2000, also finally came out in favour of eventually ending nuclear power.

13 H. H. M. Hsiao, 'The non-profit sector in Taiwan: current state, new trend and future prospect', and 'An overview of Taiwan's 300 major foundations', Taiwan Philanthropy Information Center, www.tpic.org.tw, no date (accessed January 2005).

14 J. Lu, 'The non-profit sector in Taiwan', Taiwan Philanthropy Information Center, www.tpic.org.tw, no date (accessed January 2005).

15 Himalaya Foundation, 'Overview of Taiwan's third sector – environment', Taiwan Philanthropy Information Center, www.tpic.org.tw, dated 30 December 2002.

16 Severinghaus (1989), op. cit.

17 Taiwan Environmental Protection Union website (2002).

18 These founders of the TEPU were overwhelmingly academics and professors, including the co-author of this volume.

19 Unless otherwise noted, this section is based primarily on the 'Democratic Electoral System' section in the *Taiwan Yearbook 2004*.

20 Taiwan's political evolution has been a favourite topic of numerous studies over the decades. As examples, see: Corcuff (2002); Rubinstein (1999); Tien (1989); Wachman (1994). It is very difficult, however, to find authors who have looked at the environmental stance of parties in party platforms. Taiwan studies scholars have largely ignored this dimension of the political scene in Taiwan.

21 Spelled *Guomindang* in pinyin romanization.

22 Lee is interesting and important in many other respects. Educated under the Japanese, before the KMT came to Taiwan in 1945, and a former professor of Agricultural Economics at National Taiwan University before getting into politics through the KMT, Lee maintained strong linkages with Japan and was more fluent in Japanese than in Mandarin. After he left office as President, his true leanings began to surface and he became a strong advocate for the independence of Taiwan, siding often with the position of the DPP, yet helping to establish another political party, the Taiwan Solidarity Union (see text). As a result of his actions, he was drummed out of the KMT as a traitor to the party.

23 Ji analyses why environmentalism never emerged as a major policy agenda during the KMT era and how the KMT effectively blocked most efforts at environmentalism until their power began to wane in the 1990s (Ji, 2002).

24 This volume's co-author, Chang Ch'ang-yi, was the principal investigator and coordinator of the project, which involved more than 27 professors, experts, and foreign advisors.

25 Kuomintang (KMT) website.

26 In reality, the Pan Blue component on the KMT side is split into two parts, the so-called 'Deep Blue' consisting mostly of KMT Mainlanders (especially soldiers and party officials) and the 'Light Blue' consisting mostly of KMT members drawn from the Taiwanese ethnic community.

27 'Opposition wins majority in poll', *China Post*, 12 December 2004.

28 'Election best measure of what people want: experts', *China Post*, 12 December 2004; 'Taiwan's opposition defeats pro-independence coalition in legislative election', *China Post*, 12 December 2004.

29 'DPP's poor showing due to overconfidence', *China Post*, 13 December 2004.

30 Kim (2000).

31 The flag consists of two vertical green bars separated by a white bar on which a green island of Taiwan is portrayed; hence, Chen Shui-bian and the DPP have repeatedly talked of wanting to promote a 'Green Taiwan'. Whether or not this flag would actually become the national flag of Taiwan were the island to achieve independence as a sovereign country is another matter altogether.

32 Democratic Progressive Party website.

33 'Taiwan to become "Green silicon island", Premier Yu predicts', *Taiwan News*, 20 November 2003.

34 'Activists slam DPP for 'lack of spirit', *Taipei Times*, 4 June 2004.

35 'Lee urges TSU to back Taiwanization', *China Post*, 10 February 2005.

36 Green Party Taiwan website.

37 Taiwan's situation bears striking resemblance to the polarization between Republicans and Democrats in the US.

38 The Chen government actively courted the aborigines' political support. For example, an aborigine was recruited to head the Council for Aboriginal Affairs under the Executive Yuan (formerly held by a Pan-Blue member). Chen also started a school, the Kaidagalan (named after the Pingpu tribe), to train personnel (including aborigines). A street in Taipei also was renamed with the same name as the school.

39 This section is based in part on the landmark *Taiwan 2000* study (1989), p. 736–45.

40 Primary sources for this section: *Taiwan Yearbook 2003* and *2004*.

41 The Executive Yuan is similar in structure and functions to the Executive Branch of the US government, i.e., responsible for executing laws and policies, or proposing such for passage by the Legislative Yuan.

42 Including the co-author of this text (Chang Ch'ang-yi).

43 This organization began as the Joint Commission on Rural Reconstruction (JCRR) back in the 1950s, when the US was providing major development assistance to

Taiwan, beginning with efforts to help Taiwan carry out land reform and agricultural revitalization after the Japanese colonial era.

44 The TPG was located at Chung Hsing New Town, built outside Taichung in central Taiwan expressly for this purpose. Established in 1947, it eventually consisted of an elaborate structure of agencies and units ostensibly designed to administer affairs inside Taiwan 'province', with some 124,000 employees at its height in the late 1990s, while the 'national' ROC government administered from the 'temporary national capital' of Taipei. However, given that the ROC government had only the one province of Taiwan to administer, this structure obviously resulted in considerable duplication and overlapping of responsibilities, including the environment. Thus, legislation was finally passed to formally end the post of Taiwan Governor and the Taiwan Provincial Assembly in December 1998, and dispersal of various provincial government offices and duties to other existing ROC units, as the provincial government was to be gradually phased out of existence. Political reality thus was another consequence of the reform era that started in 1987 (*ROC 1998 Yearbook*, pp. 98–9.) In 2006 some DPP lawmakers were even trying to push a bill to refloat a previously proposed plan for relocating the capital out of Taipei to central or southern Taiwan, ostensibly so that resources could be more evenly distributed and the island's economic development better balanced ('DPP lawmakers refloat capital relocation plan', *Taiwan News*, 11 December 2006.) Implicit, also, of course was the political motive of further distancing Taiwan from the ROC/KMT era, in which Taipei reigned supreme as the Mainlander stronghold.

45 P. Liu, 'Re-organizing the Executive Yuan', *Taiwan Business Topics*, April 2006, pp. 34–6.

46 An analogy in the US was the creation after 9–11 of the super-sized Department of Homeland Security, which is designed ostensibly to streamline intelligence gathering and sharing of information. Unproven, yet, is whether amalgamation of many agencies does, in fact, improve efficiency.

47 Unless otherwise noted, the data for this section were obtained from the TEPA official website: www.epa.gov.tw.

48 Interview with Professor Pen-chi Chiang (Environmental Engineering, National Taiwan University), October 2001.

49 In marked contrast to the US, where the heads of the EPA there have tended to be purely political appointees without any real environmental experience or expertise.

50 Nonetheless, there still was extensive damage to coral in the Lungkeng Ecological Reserve, as well as water pollution and disruption of fishing in the area. Taiwan eventually took the Norwegian owner to court in Norway, seeking damages beyond the roughly US$9 million already paid by the ship owners for clean-up, restoration, and compensation after the accident. But in 2005 the Norwegian court ruled against Taiwan ('Taiwan loses Kenting oil-spill suit', *Taipei Times*, 12 January 2005).

51 TEPA website, 2003; interview with Pen-chi Chiang (Environmental Engineering, National Taiwan University), October 2001.

52 Some have speculated that the underlying reason for Hau's resignation was that it was the right time to step down just before the 2004 election. He wanted to get out while he still had a good reputation, which could help promote his future. Hau in fact did run for Mayor of Taipei in 2006 and was elected in December, the start of a potentially promising political career. As for the Pinglin Interchange that sparked Hau's resignation from the TEPA in the first place, the TEPA went ahead in June 2006 and approved limited vehicular access to the Pinglin Interchange on the new Taipei-Yilan Freeway. The limit was set at 4,000 vehicles per day, in addition to those owned by residents of Pinglin Township. This decision was, in effect, a compromise designed to placate residents of the township (who wanted unrestricted use of the interchange, to boost local tourism in an area famous for tea plantations) and Taipei City, which wanted the ramp to be restricted to emergency vehicles only, in

order to protect the watershed of the Feitsui Reservoir, the main source of Taipei's water supply. Unspecified was how the daily vehicle limit was to be enforced ('EPA to open interchange at Pinglin', *Taiwan News*, 6 May 2006; 'EPA to check ramp-dam pollution link', *Taiwan News*, 5 May 2006).

53 'New EPA chief sworn in, vows no policy changes', *China Post*, 23 October 2003; 'EPA minister quits', *Taiwan News*, 2 October 2003.

54 Interview with Professor Yueh-hua Yu (Environmental Engineering, National Taiwan University), October 2001.

55 Complete details on the various units of the TEPA can be found at their website: www.epa.gov.tw. This section is also drawn from interviews and conversations with various TEPA officials conducted over the years by the authors.

56 C. Hsieh, 'Evolution of Taiwan's environmental bill', *Taipei Journal*, 15 June 2001.

57 F. Y. Lin, 'ROC enacts environmental legislation', *Taipei Journal*, 20 December 2002.

58 Interview with Professor Yueh-hua Yu (Environmental Engineering, National Taiwan University), October 2001.

59 V. Sheng, 'EPA pushes green-mark program', *Free China Journal*, 7 October 1994; *Taiwan Yearbook* 2005, p. 204.

60 'Green purchasing to reshape business', *China Post*, 23 October 2003.

61 'Green purchasing by public agencies wins EPA approval', *Taipei Times*, 11 December 2006, p. 2.

62 'Taiwan quick take: EPA launches competition', *Taipei Times*, 7 June 2006, p. 3.

63 Y. C. Yu, 'The practice of the environmental impact assessment system in Taiwan, Republic of China', paper presented at The IV National Congress of Sanitary and Environmental Engineering, San Salvador, El Salvador, October 2000 (unpublished record in TEPA, Bureau of Comprehensive Planning).

64 The search committee does the selection of new members, seven being elected each time, for two-year terms, in a rotational process. A member can serve a maximum of two consecutive terms. The Commission meets once a month, with a quorum of 11 members required.

65 Interview with Charles Y. C. Yu, Senior Specialist, Bureau of Comprehensive Planning, TEPA, October 2001.

66 Y. C. Cheng, 'Environment versus economics', *Free China Review*, April 1996, pp. 58–9.

67 See also: Pat Gao, 'Flushed with pride', *Taiwan Review*, January 2007, pp. 18–21.

68 W. Chou, 'Kaohsiung City's Love River brought back from the dead', *Panorama*, 23 August 2002.

69 'EPA to spend NT$1 billion in Danshuei River cleanup plan', *Taiwan News*, 22 August 2005.

70 'Taiwan to invest NT$80b in flood prevention', *Taiwan News*, 15 March 2005.

71 Another example of the abominable romanization of place names in Taiwan today, with the Tongyong Pinyin system.

72 'Legislature urged to support budget for flood control', *Taiwan News*, 29 June 2005; 'Cabinet touts end to flooding nightmare', *Taipei Times*, 7 August 2005, p. 2; 'Premier Yu finds route to inspection blocked by rockfall', *Taipei Times*, 26 October 2004, p. 3. For detailed background on the Keelung River and the project, see: 'Taming a river', *Taipei Review*, June 2002, pp. 12–17.

73 L. Curry, 'Taiwan's perennial water woes', *Taiwan Business Topics*, April 2006, pp. 27–8.

74 'Water quality deteriorating in key reservoirs due to farming practices', *Taiwan News*, 7 April 2006.

75 'Water problems plague Taoyuan', *Taiwan News*, 12 August 2005.

76 J. Tsai, 'Taoyuan, Taipei county residents incensed as officials grapple with sediment problem', *Taiwan Journal*, 10 March 2006.

77 'Researcher urges government to utilize rainwater efficiently', *Taiwan News*, 22 June 2004.
78 'Water at five beaches found unsuitable for swimming', *Taiwan News*, 27 June 2005.
79 'The price of water is just too low', *Taipei Times*, 12 August 2005.
80 *Taiwan Yearbook* 2005.
81 H. W. Chen and F. H. Fang, 'Air pollution control fee: the Taiwan experience', TEPA website 2001.
82 The bicycle is actually making something of a comeback in Taiwan, after decades of decline in usage, due to the heavy motor vehicle traffic and bad air. Now government support for bicycle lanes and changing public attitudes are giving bicycles something of a renaissance. The impact is probably greatest in smaller towns, rather than in the big metropolises.
83 'New EPA initiative aims to cut auto emission levels in 3 cities', *Taiwan News*, 6 January 2006.
84 See also: Jim Hwang, 'A trashy business', *Taiwan Review*, January 2007, pp. 12–17, for an interesting overview of the recycling programme in Taiwan.
85 'Food waste to join paper and plastic as material to be recycled in Taiwan', *Taiwan Journal*, 15 October 2004; 'Garbage classification project shows encouraging results, agency reports', *Taiwan News*, 3 July 2005; 'New EPA policies set to begin in January', *Taiwan News*, 5 December 2005; 'Penalties back up government's plan to cut trash', *Taipei Times*, 19 September 2005. Because of the growing bird flu scare throughout Asia, in 2006 the government decided to at least temporarily suspend use of recycled food scraps as pig fodder, and divert the food waste to production of organic fertilizer, against the protests of the swine industry, which saw the food scraps as a cheap source of feed for their pigs. How this dispute will eventually be resolved remains to be seen ('EPA to track garbage trucks, enforce sorting when new environmental policies take effect', *Taiwan Journal*, 6 January 2006).
86 'Premier urges EPA to boost recycling efforts', *Taiwan News*, 8 September 2005.
87 'Garbage reduction campaign achieves goal ahead of schedule: EPA', *Taiwan News*, 13 August 2006.
88 'Recycle more, Hsieh tells battery makers', *Taipei Times*, 8 September 2005; 'Insufficient public trash cans pose problem to residents, survey finds', *Taiwan News*, 7 June 2006, p. 3.
89 'EPA outlines new rule on plastics', *Taiwan News*, 15 November 2005.
90 'Trailblazing Taiwan leaves plastics behind', *Taiwan News*, 16 June 2004.
91 'EPA encourages people to carry their own chopsticks', *Taiwan Journal*, 20 December 2006.
92 'EPA to remove plastic bag ban on food stalls starting May 1', *Taiwan News*, 30 March 2006. Unfortunately, this led to further backtracking in 2006, as retailers and customers came to the realization that they were not quite ready for the inconvenience of a complete ban on plastic bags in daily life. An illustration of the reality that environmental protection sometimes involves two steps forward and one step backward.
93 Interview with H. G. Leu (Lu Hung-kuang), Director-General of Bureau of Solid Waste Management, TEPA, October 2001.
94 'EPA faces dilemmas in two incinerator construction projects', *Taiwan News*, 28 March 2006.
95 'Hsinchu residents reject incinerator project', *Taiwan News*, 12 November 2005; 'Toxic incinerator ordered to shut down and relocate', *Taiwan News*, 6 January 2006.
96 'Landfills poison water: protestors', *Taipei Times*, 28 July 2005.
97 'EPA lauds anti-dumping project success', *Taipei Times*, 9 February 2006, p. 2.

98 'EPA chief reopens controversy over 4th nuclear power plant', *Taiwan News*, 9 June 2005.
99 'EU law could hurt sales', *Taipei Times*, 1 April 2006, p. 12.
100 'Green electronics take root', *Taipei Times*, 20 June 2005.
101 'Electrical, electronics producers struggle to meet green standards', *Taiwan Journal*, 12 August 2005.
102 'Greenpeace tries for a bit of green at Computex fair', *Taipei Times*, 10 June 2006, p. 12.
103 E. Simon, 'E-junk pileup: tech recycling programs still in infancy', *Lansing State Journal*, 7 December 2004.
104 K. Mayfield, 'E-waste: dark side of digital age', *Wired News*, 20 January 2003 (www.wired.com/news); J. Drucker, 'Old cell phones pile up by the millions', *Wall Street Journal*, 23 September 2004.
105 'EPA to combat pollution from batteries', *Taiwan News*, 22 August 2005.
106 S. C. Ho, 'Battery recycling in Taiwan', TEPA website (2006).
107 'Cabinet approves budget for construction projects', *Taipei Times*, 14 October 2004.
108 'Protest targets development plans', *Taipei Times*, 6 June 2004.
109 'Officials optimistic over Suao-to-Hualien freeway', *Taipei Times*, 14 December 2004.
110 'Planned expressway seen damaging environment', *Taiwan News*, 18 January 2007; 'East coast expressway gets on the fast track to opposition from local environmental groups', *Taiwan Journal*, 2 February 2007.
111 'Controversial expressway necessary, president says', *Taiwan News*, 22 April 2007; 'Chen's remarks on expressway heat up debate', *Taiwan News*, 23 April 2007.
112 'Editorial: to build or not to build?', *Taipei Times*, 24 April 2007. Although technically not one of the 'Ten Nightmares', the Hsuehshan Tunnel on the new Taipei-Yilan Freeway has been a controversial public works project for some time. The highway cuts travelling time between Taipei and Ilan from two hours down to 40 minutes. The Hsuehshan Tunnel goes through the mountains between the two counties and, at 12.9 km long, is the world's fifth longest tunnel. However, the tunnel was plagued with problems during its construction and the opening was repeatedly delayed. Now in full operation, however, the tunnel has indeed proved popular and greatly eased movement by road between Taipei and Yilan and on to Hualien. It is now possible to make the trip from Hualien to Taipei in just three hours by car, a remarkable improvement from years ago when that was essentially an all-day journey ('Hsuehshan tunnel set to open by June', *Taiwan News*, 9 February 2006).
113 'Tilting trains ease east line travel woes', *Taiwan Journal*, 11 May 2007.
114 'Warning issued on major construction project plans', *Taipei Times*, 29 May 2004; 'Artificial lakes get attention of ecology groups', *Taipei Times*, 21 June 2004.
115 'Who gains from building projects?', *Taipei Times*, 14 June 2004.
116 'EPA asks for revisions to reports on the Hushan Reservoir project', *Taiwan News*, 17 February 2005.
117 'Water demand vs. environmental protection', *Taipei Journal*, 1 September 2000.
118 'Tourists and environmentalists clash', *Taipei Times*, 30 May 2004; 'Cable-car plan protested by mountain enthusiasts', *Taipei Times*, 31 May 2004; 'Activists demand public debate on cable-car plan', *Taipei Times*, 18 June 2004.
119 'Ma under fire over cable car project', *Taipei Times*, 25 July 2006, p. 3. Partly in response to this criticism over the approval of cable car systems at Yangmingshan and Sun Moon Lake without EIAs, the government announced new regulations requiring EIAs for all new projects involving landfills, incinerators, crematoriums, forest recreation centres, and cable car construction projects ('EPA tries to rein in environmentally harmful projects', *Taipei Times*, 22 February 2006, p. 2)
120 This section on nature conservation is based primarily on: *Taiwan's National Parks* (2003); *Jewels of Ilha Formosa* (2001); and 'Nature Conservation' (2005).

121 The build/operate/transfer system (BOT) is a form of project financing, which relies on private enterprises to build and operate a project for a specified period of time, and charge for the use of the facilities in order to recover costs and make a profit, before ownership is transferred to the public sector. Many public works projects in Taiwan are now constructed under BOT, which reduces the risk to the government. This system is now widely used around the world, in fact.
122 'Sacred mountain to get ecological breather', *Taipei Times*, 5 January 2006.
123 Z. Cheng, 'National identity: Taiwan looks for its roots', *Taiwan Review*, March 2007, 35–41. One can observe an interesting parallel with the long-established role of Mount Fuji in Japanese culture, with pictures of Fuji often appearing through the ages and still today as an icon of that country.
124 'Nature's springtime raiment draws crowds', *Taiwan Journal*, 24 March 2006, p. 5. These huge numbers of visitors are also a further reminder of the different attitude toward crowding that typifies Chinese (Taiwanese) society, compared with the US, as discussed in Chapter 2.
125 'Officials refute claims Taroko project is illegal', *Taiwan News*, 6 May 2006.
126 The older, Wade-Giles, romanization for this place name is shown here as an illustration of the dilemma currently confronting Taiwan, as discussed earlier in the volume, with partial implementation of Tongyong Pinyin.
127 'Making of Makao National Park still underway', *Taiwan News*, 13 August 2003.
128 'First marine park established after budget approved', *Taiwan Journal*, 1 December 2006.
129 R. Fang, 'Rugged beauty marks eastern region', *Taiwan Journal*, 20 June 2003.
130 R. Fang, 'East Rift Valley National Scenic Area offers guests many outdoor activities', *Taiwan Journal*, 16 May 2003.
131 'Council plans mega project to develop eastern Taiwan', *Taipei Times*, 11 December 2006, p. 12.
132 J. Hwang, 'Energy: weighing the policy options', *Taiwan Review*, April 2006, pp. 4–11.
133 *TEPA Yearbook*, 2006 (data from Bureau of Energy, Ministry of Economic Affairs), pp. 206–7.
134 'Government touts clean energy policy', *Taipei Times*, 6 June 2005.
135 'Taiwan's energy consumption up 2.52% in first half', *Taiwan News*, 13 August 2006.
136 'Government still committed to end of nuclear power', *Taipei Times*, 11 January 2005.
137 'Taiwan to boost coal use as nuke plants are phased out', *Taiwan News*, 29 December 2004.
138 'Plans for new Taiwan Power Company thermal plant cause row', *Taiwan News*, 24 September 2005; 'Environmentalists protest coal-fired power stations', *Taipei Times*, 22 September 2005.
139 'Taiwan mulls developing wind power', *Taiwan News*, 17 August 2005; 'Taipower opens 23 new wind turbines', *Taiwan Journal*, 27 April 2007; 'Nation tilts toward using wind power', *Taiwan Journal*, 4 May 2007.
140 'Metal industry research center orchestrates wind power group', *Taiwan Journal*, 16 September 2005.
141 'Chen urges public to get in the habit of saving energy', *Taipei Times*, 15 July 2005.
142 'County commissioner wants action on waste', *Taipei Times*, 7 June 2006, p. 3.
143 'AEC says waste relocation can be sped up', *Taipei Times*, 5 January 2006.
144 'Lawmakers go on nuclear waste tour at Nevada facility', *Taipei Times*, 27 August 2005.
145 P. Gumbel, 'Re-energized in France', *Time Magazine*, 10 March 2006, A9–A10.
146 'Reactor dreams', *The Economist*, 25 February 2006, p. 38.
147 P. Gao, 'Energy: the politics of power', *Taiwan Review*, April 2006, pp. 18–21.

148 'Cabinet vows to cut back on greenhouse emissions', *Taipei Times*, 17 February 2005.
149 'Government moves to map out energy plan', *Taipei Times*, 20 June 2005.
150 'Don't despair: most of the news on the climate front is bad, but not all of it', *The Economist*, 10 December 2005, p. 11.
151 'Ministry of Economic Affairs, Taipower set higher target for reducing emissions', *Taiwan Journal*, 6 February 2006. Shih (2005) presents an interesting analysis of Taiwan's participation in multilateral environmental agreements (MEAs) and suggests that Taiwan's involvement in such agreements has been based less on altruistic intentions and more on self-preservation and survival.
152 'Taiwan wants in on U.N. emissions talks', *Taiwan News*, 17 February 2005.
153 'EPA press conference in response to Kyoto Protocol', *China Post*, 17 February 2005.
154 'National energy conference slated for June', *China Post*, 17 February 2005.
155 Within Asia, a notable example of how difficult it will be to get beyond the Kyoto Protocol, with all its flaws, was the formation in 2006 of the so-called Asia–Pacific Partnership on Clean Development and Climate, whose members include the US, Australia, China, India, Japan, and South Korea. Ostensibly set up to complement the Kyoto Protocol and other efforts, critics have dismissed the organization as a smokescreen by some of the biggest offenders in GHG emissions. Japan is the only member that is currently trying to cut down its emissions. Moreover this Partnership's charter explicitly states that commitments are not legally binding but merely goals to aim for ('More hot air', *The Economist*, 14 January 2006, p. 46).
156 'Treat emission reduction practically, legislators say', *Taipei Times*, 6 December 2004.
157 'Cabinet unveils emissions bill', *Taiwan Journal*, 6 October 2006.
158 'Cell phone antenna sites seen as top environmental issue', *Taiwan News*, 12 February 2006.
159 The reason for this public attitude is due to the TEPA's not enforcing the anti-plastic bag policy uniformly and strictly, and hence the public tends to see the policy as a failure. This commonly happens with environmental policies, when strictly written laws and regulations run into political and economic realities.
160 In this regard, the government recently took a step forward when the TEPA launched in early 2006 a Chinese-language online clearing house for news and reports compiled from central and local government agencies on environmental conditions in residential areas all over Taiwan. The endeavour actually began in 2002 and is called the 'Residential Environmental Information System', providing information on 389 townships and cities islandwide. The website is: http://edb.epa.gov/tw/living/index.asp.

4 Whose land is it? Land use issues

1 Unless otherwise cited, this section is derived from: Williams *et al.* (1983).
2 S. R. Severinghaus, 'The emergence of an environmental consciousness in Taiwan', paper presented at 1989 Annual Meeting of the Association for Asian Studies, Washington, DC.
3 Hsieh (1964), p. 69.
4 As was true of building a harbour at Taichung, an idea the Japanese abandoned as impractical. The ROC government, however, went ahead with its Taichung Harbour project, and critics now widely deride that project as a white elephant. By contrast, regarded as one of the great positive legacies of Japanese rule is the Wushantou Reservoir in Tainan County and the contiguous Chiayi-Tainan (Chianan) Canal, which opened up much farmland to irrigation in that area. The designer of those structures, a Japanese engineer named Yoichi Hatta, is still honored by a statue of him

in front of the reservoir ('Decisions on land use must be responsible', *Taipei Times*, 15 July 2004).

5 'Atayal aboriginal cultural, historical sites, and highway history', Taroko National Park, 2004: www.taroko.gov.tw/.

6 'Atayal aboriginal', ibid.

7 The area is sometimes colloquially referred to as the 'Lishan Area' because of the dominance of the town of Lishan as the focal point of the whole basin, and where the bulk of the resident population lives.

8 'Verdict on cross-island highway: highlands need a chance to heal', *Taiwan Journal*, 13 August 2006, p. 3.

9 VACRS still has over 100 subsidiary organizations under it.

10 VACRS even built a vacation retreat for Chiang Kai-shek at Tienchih, high on a mountain peak above Fushou Shan (FSS), plus another home for his use right at FSS. The structures still stand today as tourist attractions.

11 There were obvious similarities to the way the US government has attempted to deal with its own 'aborigines', the native Americans or 'Indians' on their reservations.

12 Taiwan was not alone in this regard, as the growing intensity and frequency of serious storms around the world seemed to support the arguments of critics of global warming and their urging heightened control of GHG ('Asian typhoons take unusual turn in worst season on record', *China Post*, 13 September 2004).

13 'Disaster prevention must improve', *Taipei Times*, 15 July 2004, p. 8.

14 'Floods prompt review of land-use policy', *Taiwan Journal*, 30 July 2004, p. 1.

15 'Rebuilt plant will be flood proof', *Taipei Times*, 5 October 2004.

16 'Government rethinks environmental policies', *Taiwan Journal*, 30 July 2004, p. 1.

17 'President Chen urges end to trading of barbs over typhoon', *China Post*, 30 August 2004; 'Environment needs official action, not resignations', *Taiwan News*, 10 September 2004.

18 'President Chen', ibid.

19 'Academics urge tackling environmental problems', *Taipei Times*, 16 January 2005, p. 4.

20 J. Baum, 'Herb hits hard: typhoon highlights inadequate protection measures', *Far Eastern Economic Review*, 22 August 1996: 24.

21 Hsu (2003) analyses the effects of democratization and changing inter-governmental relations on Taiwan's watershed management, especially in the Taipei Water Special Area, and observes that effective watershed management in Taiwan must deal with local participation, intergovernmental coordination, and conflicting demands on water and land resources management.

22 'Aboriginals scapegoated, Chin says', *Taipei Times*, 13 July 2004.

23 'Verdict on cross-island highway: highlands need a chance to heal', *Taiwan Journal*, 13 August 2004, p. 3; 'Cabinet to restrict storm repairs in sensitive areas', *Taipei Times*, 13 July 2004, p. 1.

24 'Taiwan points way with "Green Front" for land recovery', *Taiwan News*, 21 January 2005.

25 'Cabinet approves mountain land development plan', *Taiwan News*, 20 January 2005.

26 'Cabinet approves changes in land act', *Taiwan News*, 26 May 2005.

27 'Ecotechnology restores ravaged landscape', *Taiwan Journal*, 9 July 2004; 'New ways to stop erosion work: report', *Taipei Times*, 21 September 2004.

28 'Defending ecological engineering', *Taipei Times*, 13 July 2004.

29 'Symposium aims to balance development with environment', *Taiwan Journal*, 17 November 2006.

30 'Forest cops target poaching, illegal logging', *Taiwan Journal*, 16 July 2004.

31 'Protesters slam road repair ban', *Taipei Times*, 16 November 2004; 'Aboriginal protesters scuffle with police', *Taiwan News*, 16 November 2004; 'Aboriginal officials criticize draft land restoration bill', *Taipei Times*, 3 February 2005.

32 'Government's land policies irrational, farmers claim', *Taiwan News*, 21 January 2005.

33 As this book was about to go to press, the government officially approved a 13th aboriginal group, the Sakizaya (or Sakiraya), a small tribe estimated at 5,000–10,000 who had previously been lumped together with the Ami since the Japanese colonial era. Living mostly in Hualien City and contiguous areas of Hualien County, the Sakizaya are the latest case of aboriginal groups struggling to re-establish their identity and place in modern Taiwan ('Sakizaya becomes the 13th indigenous group', *Taiwan Journal*, 26 January 2007; also see: Z. Cheng, 'Ingenious people: the secret's out', *Taiwan Review*, April 2007, 40–5).

34 As evidence of the growing power of the aborigines, large amounts of Aboriginal Reserved Land are now being sold to aborigines at low prices (but the amount is not publicly known). The Aboriginal Council budget has never been cut (unlike other government agencies), and the Council wields significant power. As of mid-2006, 10 out of 235 legislators were aborigines, of whom five were in the Pan Blue camp, and three in the Pan Green, the remaining two independent.

35 'Aboriginals scapegoated, Chin says', *Taipei Times*, 13 July 2004.

36 'Government rethinks environmental policies', *Taiwan Journal*, 30 July 2004, p. 1.

37 'Tribes need more say in relocation', *Taipei Times*, 23 July 2004.

38 See, for example: Chang and Huang (2001).

39 See Chang (1992).

40 In the autumn of 2006, the authors of this volume revisited Fushou Shan (FSS) and Wuling Farms and saw dramatic changes from the past. Both farms have significantly reduced fruit cultivation, as it is no longer profitable. Vegetable cultivation continues, however, especially at FSS. Trucks laden with recently harvested cabbages could be frequently encountered on the roads to Ilan and Hualien. Wuling has gone heavily into tourism, benefiting from its location on the margin of Shei-pa National Park. Various hotels and tourist facilities there have made Wuling a key tourist attraction in the upper Tachia basin, and a jumping-off point for backpackers and mountain climbers heading westward into the park.

41 The number of retired mainland soldier/farmers is also diminishing, as they gradually age and die off, and hence there is a reduced need now to find employment for these people.

42 The VACRS' Chinese-style hotel, once the 'Queen' of the lodgings available at Lishan, now stands abandoned and boarded up, slowly decaying away as a symbol of a dying era in the mountains. Yet, tourists flock to the newer and more luxurious facilities at Wuling.

43 'Farmers face challenges of WTO membership', *Taipei Journal*, 11 January 2002, p. 8.

44 'Taiwan agriculture facing difficult restructuring', *Taipei Journal*, 17 May 2002, p. 8.

45 'Free trade in agriculture leads to concern over domestic food supply', *Taipei Journal*, 16 August 2002.

46 'Free trade', ibid.

47 'Taiwan agriculture facing difficult restructuring', *Taipei Journal*, 17 May 2002, p. 8.

48 'COA announces top ten rural villages', *Taiwan Journal*, 30 March 2007.

49 O. Chung, 'Fruit for the thirsty', *Taiwan Review*, January 2004, pp. 18–23; P. Gao, 'A fruitful future', *Taiwan Review*, January 2004, pp. 4–11; J. Hwang, 'Growing the competitive edge', *Taiwan Review*, January 2004, pp. 12–17; L. Kao, 'Fruit from the swallow's nest', *Taiwan Review*, January 2004, pp. 24–9.

50 This section on organic farming is based particularly on: 'Organic farming develops following in Taiwan', *Taiwan Journal*, 16 May 2003, p. 8; and 'Hualien government vows to push organic agriculture as core policy', *Taiwan Journal*, 24 December 2004, p. 3.

51 S. Chuang (2006) 'Nutrition: increasingly organic', *Taiwan Review*, May 2006, pp. 34–9.
52 'Policy: targets for fiscal year 2005', Council of Agriculture: http://eng.coa.gov.tw.
53 See, for example: Chang (1996); Tsai *et al.* (2006a).
54 'Government to close public wells to control sinking of land', *Taiwan News*, 23 April 2005.
55 'Farmland water conservancy facilities help protect the soil', *Taiwan News*, 3 November 2005.
56 'Reservoir plan seen as harmful to environment', *Taiwan News*, 4 May 2006.

5 Reflections: toward a green silicon island

1 For another cautiously optimistic overview of Taiwan's environmental progress, see: Oscar Chung, 'Clearing a way', *Taiwan Review*, January 2007, pp. 4–8.
2 Chang and Chen (2004, 2005).
3 The water may be drinkable as it leaves the water purification plants, but the piping and storage system (often in large tanks on top of buildings) is where re-contamination of the water typically takes place.
4 The April 2006 issue of the journal *Taiwan Business Topics*, published by the American Chamber of Commerce in Taiwan, had an excellent analysis and presentation of 'Taiwan's perennial water woes'.
5 Minamata is the infamous site in Japan where consumption of seafood contaminated by mercury dumped into the bay by a chemical company over several decades led by the early 1950s to the deaths of over 700 people and the crippling of another 9,000, by some estimates. The medical problem became known as 'minamata disease' and was a key issue in promoting the environmental movement in Japan. See: Karan (2005), p. 359ff.
6 As quoted in: www.state.nj.us/dep/dsr/sustainable-state.
7 Sustainability has even begun to become fashionable in wealthier countries. See, for example: 'Green living: is sustainability the new luxury?', *Time Style & Design*, Summer, 2006.
8 UN World Commission on Environment and Development (1987). Uhl (2004) presents a detailed analysis of how nations and societies everywhere could move in the direction of a sustainable world.
9 Also see: Chang and Tsai (2002)
10 To be fair, the Chen government did not originate the idea of sustainable development in Taiwan. The concept has been floating around since at least the early 1990s in Taiwan among various groups and organizations. For example, in 1995 the Taiwan Business Council for Sustainable Development was created with 13 founding companies in concert with the Earthplace Foundation. The Council was expanded and refocused in 1997 with 31 member companies. The chairman at that time was also a member of the government's National Council for Sustainable Development. The group's goals are basically to get private businesses on board the drive to sustainability and eco-efficient practices (www.bcsd.org.tw).
11 'Taiwan needs civic movement for sustainability', *Taiwan News*, 9 June 2006, p. 9.
12 Kelly Her, 'Changing roles, a high-tech adventure', *Taiwan Review*, March 2004, pp. 4–11.
13 Wu Rong-I and Cheng Cheng-Mount, 'Economics: going up', *Taiwan Review*, June 2002, pp. 30–5.
14 Jim Hwang, 'Industry: structural reengineering', *Taiwan Review*, July 2002, pp. 4–11.
15 Wu and Cheng, 2002, op. cit.
16 There are some who feel that Chen Shui-bian may be an example of someone rising beyond the level of their capabilities. They remember the outstanding job he did as Mayor of Taipei in the mid-1990s, when he came onto the political stage and brought

dramatic improvements to Taipei with bold and imaginative policies, especially in transportation, and a dedication to principles and integrity that won widespread support (see: 'For Taipei mayor, high year-end score', *Free China Journal*, 12 January 1996, p. 7). Perhaps Chen was never meant to be President, a vastly more difficult job, and for which history may not be kind in its judgment of his tenure

17 President Chen vowed in late July 2006 to pay more attention to the environment, seemingly trying to recapture some of the popular support he once had ('President vows to better watch over ecology', *Taiwan News*, 26 July 2006).

18 'Ecological integrity is a strategic priority', *Taiwan News*, 8 June 2005.

19 'Challenge 2008: the Six-Year National Development Plan', Government Information Office, 21 May 2002 (www.gio.gov.tw/taiwan-website/4-oa/20020521/2002052101.html). Also see: Don Shapiro, 'Rising to the challenge', *Taiwan Review*, August 2002, pp. 24–31.

20 Including the co-author of this book, Chang Ch'ang-yi, who attended as a member of the National Council for Sustainable Development. See: National Council for Sustainable Development (2006).

21 'Green development conference threatened by environmentalists', *Taipei Times*, 21 April 2006, p. 3.

22 'Environmentalists plan to confront government', *Taiwan News*, 21 April 2006.

23 'Conferees debate road map to responsible development', *Taiwan Journal*, 21 April 2006, p. 3.

24 Chan and Spengler, two professors of environmental health (National Taiwan University and Harvard, respectively), in the late 1990s laid out detailed recommendations for a strategic plan that the TEPA should develop if it wished to achieve sustainable development in Taiwan. Their concluding recommendation is worth quoting: 'For a sustainable Taiwan there needs to be philosophical grounding in cultural norms and religious principles. The TEPA can advance broad-based participation among cultural, religious, social, political, and economic leaders towards defining "ecology" in the context of Chinese and Western culture.' In other words, they reinforced the importance, as stated early in this chapter, of a holistic integration of political, cultural, and social systems in a sustainable economy (Chang-Chuan J. Chan and John D. Spengler, 'Creating environmental value in Taiwan', July 1998, strategic planning for the TEPA).

25 'Conference seeks to address question of nuclear power', *Taipei Times*, 23 April 2006, p. 2. After the conference, at further meetings in May, agreement was reached on some of the unresolved issues, bringing the total consensus number to 267.

26 'President urges nation to embrace sustainability plan', *Taiwan News*, 23 April 2006.

27 'Democratic Progressive Party insists on nuclear-free Taiwan dream', *Taiwan News*, 28 April 2006; 'Yu says DPP should discuss power issue', *Taipei Times*, 28 April 2006, p. 3.

28 'Legislation on moving nuclear waste approved', *Taipei Times*, 29 April 2006, p. 3.

29 'Academics cast doubt on economic meeting's outcome', *Taipei Times*, 17 July 2006, p. 3.

30 'Environmental groups quit economic confab', *Taiwan News*, 28 July 2006.

31 'Economic conference generates much disillusion', *Taipei Times*, 29 July 2006, p. 12.

32 'Government upbeat, industries disappointed at results', *Taipei Times*, 29 July 2006, p. 3; 'ESC dominated by clashes over China policy', *Taiwan News*, 28 July 2006.

33 'Conference on sustaining Taiwan's economic development; industry panel sets high economic growth targets', *Taipei Times*, 28 July 2006, p. 2.

34 'Conference on sustaining Taiwan's economic development: social group proposes national pension law', *Taipei Times*, 28 July 2006, p. 2.

35 'Lee asks Chen for less hot air, better policies', *Taiwan News*, 30 July 2006.

36 See: 'Sustainable economy must place nation first', *Taipei Times*, 31 July 2006, p. 8; 'Economic policy should meet people's needs', *Taiwan News*, 26 July 2006.

37 The extent of those linkages are impressively summarized in: 'Taiwan and China: too close for comfort?', *Asia Times*, 17 May 2007.

38 See, for example: 'Poll says majority wants say in Taiwan sovereignty issues', *Taiwan News*, 16 March 2007.

39 L. Kao, 'High-tech concentration', *Taiwan Review*, March 2004, pp. 12–17.

40 'Environmental science and technology parks in Taiwan', TEPA webpage (www.epa.gov.tw/english) (pollution control); 'EPA vows to establish resource recycling society', *Taiwan News*, 11 May 2004.

41 'Taiwan bucks the green trend', *Taipei Times*, 11 February 2005, p. 2.

42 'Kuokuang Petrochemical agrees to swap shares for land', *Taiwan Journal*, 19 May 2006; 'Six shareholders finalize deal to build Yunlin Petrochemical Technology Park', *Taipei Times*, 5 February 2005, p. 10.

43 'EPA voices doubts about new steel mill', *Taiwan Journal*, 26 May 2006, p. 3.

44 'Taiwan bucks', op. cit.

45 The American Chamber of Commerce in Taiwan publishes an annual White Paper on Taiwan that carefully analyses the current strengths and weaknesses of Taiwan's business environment, especially from the perspective of foreign investors. The report always includes a section on the environment and the AmCham's recommendations for improving environmental policies and programmes. The government pays attention to this. See: *Taiwan Business Topics*, May 2006 (V. 36, No. 5) for the '2006 Taiwan White Paper' issue. Also see: *Taiwan Business Topics*, June 2005 (V. 35, No. 2) for a special issue focus on industry and its relationship with the environment.

46 For example, the AmCham's 2002 White Paper had this pithy summation of Taiwan: 'Taiwan's image as a hi-tech manufacturing center contrasts sharply with the reality of its inadequate infrastructure. The island needs to upgrade power and water supplies, transportation and telecommunication facilities, industrial waste disposal and sewage treatment sites, and other infrastructure essential to quality of life and business efficiency' ('Taiwan White Paper', *Taiwan Business Topics*, May 2002, p. 10).

47 'To remain competitive, nation needs to be unified', *Taipei Times*, 25 October 2004, p. 8.

48 'Growth competitiveness index rankings: 2005 and 2004 comparisons', World Economic Forum website (www.weforum.org/site).

49 'Is Taiwan the sick tiger of Asia?', *Taipei Times*, 6 June 2005, p. 8; 'Taiwan losing ground to South Korea, analysts say', *Taipei Times*, 7 April 2006, p. 12.

50 See, for example: Jacques deLisle, 'Taiwan's democracy and lessons from yet another election', *Foreign Policy Research Institute E-Notes*, 16 December 2005 (www.fpri.org.).

51 In actuality, some 76 variables of great variety are used in calculating the index. See: National Council for Sustainable Development (2006).

Bibliography

(Note: The following bibliography contains all the books, government publications, and journal articles used in this volume and cited in the endnotes. The full citation information is given below, *except for* newspapers, news magazines, and website articles, which are too voluminous in number and hence cited in full only in the endnotes, a practice increasingly common in social science publications.)

Agricultural Statistics Yearbook (2004) Taipei: Executive Yuan, Council of Agriculture.

Ahern, E. M. and Gates, H. (1981) *The Anthropology of Taiwanese Society*, Stanford, CA: Stanford University Press.

Allee, M. A. (1994) *Law and Local Society in Late Imperial China: Northern Taiwan in the Nineteenth Century*, Stanford, CA: Stanford University Press.

Arrigo, L. G. (1994) 'The environmental nightmare of the economic miracle: land abuse and land struggles in Taiwan', *Bulletin of Concerned Asian Scholars* 26 (1–2): 21–44.

Arrigo, L. G., Lin, T. L. and Lin, Y. M. (1996) 'Environmental conditions and environmental law in Taiwan', in A. Capria *et al.* (eds) *A World Survey of Environmental Law*, Milan, Italy: Giuffre Editore, 765–77.

Bain, I. (1993) *Agricultural Reform in Taiwan: From Here to Modernity?* Hong Kong: The Chinese University Press.

Bello, W. and Rosenfeld, S. (1990) *Dragons in Distress: Asia's Miracle Economies in Crisis*, San Francisco, CA: Institute for Food and Development Policy.

Berger, S. and Lester, R. (2005) *Global Taiwan: Building Competitive Strengths in a New International Economy*, Armonk, NY: M. E. Sharpe.

Bigio, A. G. and Dahiya, B. (2004) *Urban Environment and Infrastructure: Toward Livable Cities*, Washington, DC: World Bank.

Bing, S. (1986) *Taiwan's 400 Year History*, Taipei: Taiwanese Cultural Grass Roots Association.

Booth, D. E. (2004) *Hooked on Growth: Economic Addictions and the Environment*, Lanham, MD: Rowman and Littlefield Publishers.

Calkins, R., Chern, W. S. and Tuan, F. C. (eds) (1992) *Rural Development in Taiwan and Mainland China*, Boulder, CO: Westview Press.

Challenge 2008: The Six-Year National Development Plan (2002), Taipei: Government Information Office, www.gio.gov.tw/taiwan-website/.

Chang, C. Y. (1992) 'A study of environmental stress and political conflict over slopeland development in Taiwan', *International Journal of Anthropology-Global Bioethics*, (5) 1: 43–50.

—— (1996) 'Environmental impact of resource use activities on coastal zones in

Taiwan', in R. B. Singh and R. Majoral (eds) *Development Issues in Marginal Regions*, London: Oxford and IBH Publishing Co PVT Ltd, 305–9.

Chang, C. Y. and Chen, W. M. (2004) 'The spatial analysis of the effectiveness of environmental policy implementation: a case of paying for garbage disposal by the bag in Taipei City', *Hwa Kang Geographical Journal* (17): 155–78.

—— (2005) 'The cost-benefit analysis of garbage pricing by the bag policy in Taipei City', *Journal of the Land Bank of Taiwan*, 42 (1): 115–29.

Chang, C. Y. and Huang, Y. W. (2001) 'Interaction between state and society: the establishment of national parks and the awareness of autonomy of the Aborigines', *Journal of Geographical Science* (30): 1–18.

Chang, C. Y. and Liang, B. K. (2005) 'Cultural economy and social capital of aboriginal tribe tourism: a case study of the Saviki', *Journal of Geographical Science* (39): 31–51.

Chang, C. Y. and Lin, C. C. (2006) 'Applying public participation geographic information system to the study of the traditional territories of indigenous peoples: a case study of the Atayal Smangus community', *Journal of Geographical Science* (41): 65–82.

Chang, C. Y. and Tsai, B. W. (2002) 'Sustainable development and land use changes in Taiwan', *Geo-Information Science*, 4 (1): 71–3.

Chen, D. W. (1994) 'The emergence of an environmental consciousness in Taiwan', in M. A. Rubinstein (ed.) *The Other Taiwan*, 257–86.

Chen, E. K. Y., Williams, J. F. and Wong, J. (eds) (1991) *Taiwan: Economy, Society and History*, Hong Kong: University of Hong Kong, Centre of Asian Studies.

Clough, R. N. (1978) *Island China*, Cambridge: Harvard University Press.

Cooper, C. (2003) 'Quietly sowing the seeds of activism', *Far Eastern Economic Review*, 10 April.

Corcuff, S. (ed.) (2002) *Memories of the Future: National Identity Issues and the Search for a New Taiwan*, Armonk, NY: M. E. Sharpe.

Davidson, J. W. (1903) *The Island of Formosa, Past and Present*, London: Macmillan.

De Blij, H. J., Muller, P. O. and Williams, R. S. (2004) *Physical Geography of the Global Environment*, New York: Oxford University Press.

Dyson, T., Casen, R. and Visaria, L. (eds) (2004) *Twenty-First Century India: Population, Economy, Human Development, and the Environment*, New York: Oxford University Press.

Economy, E. C. (2004) *The River Runs Black: The Environmental Challenge to China's Future*, Ithaca, NY: Cornell University Press.

Goeldner, C. R. and Ritchie, J. R. B. (2003) *Tourism, Principles, Practices, Philosophies*, New York: John Wiley and Sons, Inc., 9th edn.

Gold, T. B. (1986) *State and Society in the Taiwan Miracle*, Armonk, NY: M. E. Sharpe.

Gordon, B., Mackay, R. and Rehfuess, E. (2004) *Inheriting the World: The Atlas of Children's Health and the Environment*, Geneva: World Health Organization.

Harris, F. (ed.) (2004) *Global Environmental Issues*, Hoboken, NJ: Wiley.

He, B. C. (1991) *China on the Edge: The Crisis of Ecology and Development*, San Francisco, CA: China Books and Periodicals.

Ho, P. (2001) 'Greening without conflict? Environmentalism, NGOs and civil society in China', *Development and Change*, 32: 893–921.

Ho, S. P. S. (1978) *Economic Development of Taiwan, 1860–1970*, New Haven, CT: Yale University Press.

Hsiao, M. H. H. (1989) 'Changing public environmental perceptions, attitudes and behaviors', in *Taiwan 2000: Balancing Economic Growth and Environmental Protection*, Taipei: Academic Sinica, Institute of Ethnology, 369–76.

—— (1999) 'Environmental movements in Taiwan', in Y. S. Lee and A. Y. So (eds), *Asia's Environmental Movements: Comparative Perspectives*, Armonk, NY and London: M. E. Sharpe, 31–54.

Hsiao, M. H. H. and Tseng, H. P. (1999) 'The formation of environmental consciousness in Taiwan: intellectuals, media, and the public mind', *Asian Geographer* 18 (1–2): 99–109.

Hsiao, M. H. H., Cheng, W. Y. and Chan, H. S. (eds) (1989) *Taiwan: A Newly Industrialized State*, Taipei: National Taiwan University.

Hsiao, M. H. H., Chiang, P. C., Severinghaus, L. L. and Chu, Y. P. (1993) *Taiwan 2000: Balancing Economic Growth and Environmental Protection*, Taipei: Commonwealth Publishing Co. Ltd (Chinese-language version of *Taiwan 2000* study).

Hsieh, C. M. (1964) *Taiwan–Ilha Formosa: A Geography in Perspective*, Washington: Butterworths.

Hsiung, J. C. *et al.* (eds) (1981) *The Taiwan Experience, 1950–1980*, New York: Praeger Publishers.

Hsu, J. Y. (1995) 'The struggle over local space: a case study of the environmental movement in Taiwan', *CNS* 6 (1): 113–24.

Hsu, S. H. (2003) 'Democratization, intergovernmental relations, and watershed management in Taiwan', *Journal of Environment and Development*, 12 (4): 455–63.

Huddle, N., Reich, M. and Stiskin, N. (1975) *Island of Dreams: Environmental Crisis in Japan*, New York: Autumn Press.

Jasanoff, S. and Martello, M. L. (eds) (2004) *Earthly Politics: Local and Global in Environmental Governance*, Cambridge, MA: MIT Press.

Jewels of Ilha Formosa (2001) Taipei: Ministry of the Interior, Construction and Planning Administration.

Ji, S. J. (2002) *Environmental Politics in Taiwan*, East Lansing, MI: Michigan State University, unpublished doctoral dissertation.

Kao, S. F. (2002) *Risk Perceptions and Environmental Mobilization – Tracking the Transformation of Collective Actions in a Radiation Contamination Incident in Taiwan*, East Lansing, MI: Michigan State University, unpublished doctoral dissertation.

Karan, P. P. (2005) *Japan in the 21st Century: Environment, Economy, and Society*, Lexington, MA: University Press of Kentucky.

Kerr, G. H. (1965) *Formosa Betrayed*, Boston, MA: Houghton Mifflin.

—— (1974) *Formosa: Licensed Revolution and the Home Rule Movement, 1895–1945*, Honolulu: The University of Hawaii Press.

Kim, S. (2000) 'Democratization and environmentalism: South Korea and Taiwan in comparative perspective', *Journal of Asian African Affairs*, XXXV (3): 287–302.

Knapp, R. G. (1980) *China's Island Frontier: Studies in the Historical Geography of Taiwan*, Honolulu: The University Press of Hawaii.

Kuo, S. W. Y., Ranis, G. and Fei, J. C. H. (1981) *The Taiwan Success Story: Rapid Growth with Improved Distribution in the Republic of China, 1952–1979*, Boulder, CO: Westview Press.

Lai, T. H., Myers, R. H. and Wou, W. (eds) (1991) *A Tragic Beginning: The Taiwan Uprising of February 28, 1947*, Stanford, CA: Stanford University Press.

Lawrence, S. (2003) 'NGO's long, hard road to win official approval', *Far Eastern Economic Review*, 10 April.

Lee, T. H. (1999) *The Road to Democracy: Taiwan's Pursuit of Identity*, Tokyo: PHP Institute, Inc.

Leng, T. K. (2005) 'Sustainable development in Canada and Taiwan: comparative and international perspectives', in P. G. Harris (ed.) *Confronting Environmental Change in East and Southeast Asia: Eco-Politics, Foreign Policy, and Sustainable Development*, London: Earthscan, 201–18.

Liu, A. P. L. (1987) *Phoenix and the Lame Lion: Modernization in Taiwan and Mainland China, 1950–1980*, Stanford, CA: Hoover Institution.

Meskill, J. M. (1979) *A Chinese Family: The Lins of Wu-feng, Taiwan, 1729–1895*, Princeton, NJ: Princeton University Press.

Myers, N. and Kent, J. (2004) *The New Consumers: The Influence of Affluence on the Environment*, Washington, DC: Island Press.

National Council for Sustainable Development (NCSD) (2006) *Annual Report on National Sustainable Development*, Taipei: Executive Yuan, ROC.

'Nature conservation' (2005) *Taiwan Yearbook 2005* (www.gov.tw/Yearbook).

Perez, O. (2004) *Ecological Sensitivity and Global Legal Pluralism: Rethinking the Trade and Environment Conflict*, Portland, OR: Hart.

Randolph, J. (2004) *Environmental Land Use Planning and Management*, Washington, DC: Island Press.

Ranis, O. (ed.) (1992) *Taiwan: From Developing to Mature Economy*, Boulder, CO: Westview Press.

Reardon-Anderson, J. (1992) *Pollution, Politics, and Foreign Investment in Taiwan: The Lukang Rebellion*, Armonk, NY: M. E. Sharpe.

Reisner, M. (1986) *Cadillac Desert: The American West and Its Disappearing Water*, New York: Penguin Books.

ROC Yearbook (1998) Taipei: Government Information Office.

Rubinstein, M. A. (ed.) (1994) *The Other Taiwan: 1945 to the Present*, Armonk, NY: M. E. Sharpe.

—— (ed.) (1999) *Taiwan: A New History*, Armonk, NY: M. E. Sharpe.

Selya, R. M. (2004) *Development and Demographic Change in Taiwan, 1945–1995*, New Jersey: World Scientific Publishing Co.

Sharpley, R. and Telfer, D. J. (eds) (2002) *Tourism and Development: Concepts and Issues*, Clevedon: Channel View Publications.

Shaw, S. L. and Williams, J. F. (1991) 'The role of transportation in Taiwan's regional development', *Transportation Quarterly*, 45 (3): 271–96.

Shepherd, J. R. (1993) *Statecraft and Political Economy on the Taiwan Frontier, 1600–1800*, Stanford, CA: Stanford University Press.

Shih, W. C. (2005), 'Taiwan's international environmental policy: balancing trade and the environment', in P. G. Harris (ed.) *Confronting Environmental Change in East and Southeast Asia: Eco-Politics, Foreign Policy, and Sustainable Development*, Tokyo: United Nations University Press: 119–32.

Simon, D. F. and Kau, M. Y. M. (eds) (1992) *Taiwan: Beyond the Economic Miracle*, Armonk, NY: M. E. Sharpe.

Sims, R. E. H. (ed.) (2004) *Bioenergy Options for a Cleaner Environment in Developed and Developing Countries*, Boston, MA: Elsevier.

Smil, V. (1984) *The Bad Earth: Environmental Degradation in China*, Armonk, NY: M. E. Sharpe.

—— (1993) *China's Environmental Crisis: An Inquiry into the Limits of National Development*, Armonk, NY: M. E. Sharpe.

—— (2004) *China's Past, China's Future: Energy, Food, Environment*, New York: Routledge.

Sorensen, A., Marcotullio, P. J. and Grant, J. (eds) (2004) *Towards Sustainable Cities: East Asia, North American and European Perspectives on Managing Urban Regions*, Burlington, VT: Ashgate.

Speth, J. G. (2004) *Red Sky at Morning: America and the Crisis of the Global Environment*, New Haven, CT: Yale University Press.

Sutton, P. W. (2004) *Nature, Environment, and Society*, New York: Palgrave Macmillan.

Taiwan Environmental Protection Administration (2004), *Environmental Yearbook*, Taipei: Executive Yuan, TEPA. (in Chinese).

—— (2006) *Yearbook of Environmental Protection Statistics, Republic of China*, Taipei: Executive Yuan, TEPA (in Chinese/English).

Taiwan 2000: Balancing Economic Growth and Environmental Protection (1989) Taipei: The Steering Committee, Taiwan 2000 Study, c/o Academia Sinica, Institute of Ethnology.

Taiwan Yearbook 2003, Taipei: Government Information Office, www.gio.gov.tw/-taiwan-website/5-gp/yearbook.

Taiwan's National Parks (2003) Taipei: Ministry of the Interior, Construction and Planning Agency.

Tang, S. Y. and Tang, C. P. (1997) 'Democratization and environmental politics in Taiwan', *Asian Survey*, 37 (3): 281–94.

Tsai, B. W. (1999) 'Democratization and the environment: entrepreneurial politics and interest representation in Taiwan', *The China Quarterly*, 158 (June): 350–66.

Tien, H. M. (1989) *The Great Transition: Political and Social Change in the Republic of China*, Stanford, CA: Hoover Institution.

Trefil, J. S. (2004) *Human Nature: A Blueprint for Managing the Earth – by People, for People*, New York: Times Books/Henry Holt.

Tsai, B. W., Chang, K. T., Chang, C. Y. and Chu, C. M. (2006) 'Analyzing spatial and temporal changes of aquaculture in Yunlin County, Taiwan', *The Professional Geographer*, 58 (2): 161–71.

Tsai, B. W., Chang, C. Y., Lin, C. C. and Lo, Y. C. (2006) 'PPGIS and indigenous society: new partnership of indigenous peoples in Taiwan', *Geography Research Forum*, 26 (4).

Uhl, C. (2004) *Developing Ecological Consciousness: Path to a Sustainable World*, Boulder, CO: Rowman and Littlefield Publishers.

UN World Commission on Environment and Development (1987) *Our Common Future*, Oxford and New York: Oxford University Press.

Wachman, A. M. (1994) *Taiwan: National Identity and Democratization*, Armonk, NY: M. E. Sharpe.

Williams, J. F. (1980) 'Sugar: the sweetener in Taiwan's development', in R. Knapp (ed.) *China's Island Frontier: Studies in the Historical Geography of Taiwan*, Honolulu: The University Press of Hawaii, 219–51.

—— (1981) 'Agricultural use of slopelands in Taiwan', in C. Pannell and C. Salter (eds) *China Geographer*, No. 11: 89–111, Boulder, CO: Westview Press.

—— (1992) 'Environmentalism in Taiwan', in D. F. Simon and M. Y. M. Kau (eds), *Taiwan: Beyond the Economic Miracle*, Armonk, NY: M. E. Sharpe, Inc., 187–210.

—— (1994) 'China and Taiwan', in D. Dwyer (ed.) *China: The Next Decades*, London: Longman Scientific and Technical, 222–42.

—— (1996) 'The quality of life in Taiwan: an environmental assessment', *American Asian Review*, XIV (3), Fall: 79–105.

—— (2003a) 'Who are the Taiwanese? Taiwan in the Chinese Diaspora', in L. J. C. Ma and C. Cartier (eds) *The Chinese Diaspora: Space, Place, Mobility, and Identity*, Boulder, CO: Rowman and Littlefield, 163–89.

—— (2003b) 'The role of secondary cities in rapidly industrializing countries: the example of Kaohsiung, Taiwan', in A. K. Dutt, A. G. Noble, G. Venugopal and S. Subbiah (eds) *Challenges to Asian Urbanization in the 21st Century*, Dordrecht: Kluwer Academic Publishers, 225–41.

Williams, J. F. and Chang, C. Y. (1989) 'Paying the price of economic development in Taiwan: environmental degradation', *Journal of Oriental Studies* (University of Hong Kong), XXVII (1–2): 59–78. Reprinted in Rubinstein, M. A. (1994), *The Other Taiwan: 1945 to the Present*, Armonk, NY: M. E. Sharpe, 237–56.

Williams, J. F., Chang, C. Y. and Wang, C. Y. (1983) 'Land settlement and development: a case study from Taiwan', *The Journal of Developing Areas*, 18 (1): 35–52.

Wisner, B. and Blaikie, P. M. (eds) (2004) *At Risk: Natural Hazards, People's Vulnerability and Disasters*, New York: Routledge.

Useful websites

Environmental Protection Administration, Taiwan: www.epa.gov.tw

Democratic Progressive Party: www.dpp.org.tw

Green Party Taiwan: www.taiwanese.com/gptaiwan

Kuomintang (KMT) Party: www.kmt.org.tw

National Council for Sustainable Development (NCSD): www.ivy2.epa.gov.tw/NSDN/

New Party (NP): www.np.org.tw

People First Party (PFP): www.pfp.org.tw

Residential Environmental Information System (TEPA): www.edb.epa.gov/tw/living/index.asp

Taiwan Environmental Protection Union (TEPU): www.tepu.org.tw

Taiwan Solidarity Union (TSU): www.tsu.org.tw

Taiwan Yearbook (latest edition), Government Information Office: www.gov.tw/Yearbook.

Taiwan newspapers and journals (in English)

The following newspapers and periodicals published in Taiwan were used extensively in this volume. During the martial law era, some of these publications did not exist, others tended to be more propaganda outlets for the government than sources of critical information. Since the late 1980s these publications have become objective, factual sources of up-to-date information about all facets of Taiwan, even though the market is limited for English-language media in Taiwan. These publications are now essential reading for anyone wanting to be informed about current developments in Taiwan. Moreover, unlike in the US, these kinds of publications in Taiwan frequently carry reports and analyses of environmental issues. Indeed, reporting about the environment in all the visual and printed media (Chinese and English) is much more common in Taiwan than in the US. Articles from these publications used as sources for this study are fully cited in the endnotes for each chapter, but are not repeated in the bibliography above because of the sheer volume of such items.

China Post (www.chinapost.com.tw): A daily online newspaper (also published in hard copy form in Taiwan). The oldest English-language newspaper in Taiwan, usually pro-blue in its political outlook, but still useful.

Taipei Times (www.taipeitimes.com): A daily online newspaper (also published in hard copy form in Taiwan), launched in 1999, one of the leading English-language newspapers in Taiwan. This paper and the *Taiwan News* (below) normally have a pro-green slant to their reporting.

Taiwan Business Topics (www.amcham.com.tw): A monthly, glossy news magazine published by the American Chamber of Commerce in Taiwan. Aimed primarily at the large expatriate business community, this news magazine often has articles focused on environmental issues, or related economic, social, and political issues, and is well-illustrated and authoritative in its facts and analyses. Its annual 'White Paper' on Taiwan is especially useful.

Taiwan Journal (www.taiwanjournal.nat.gov.tw). A weekly newspaper, established in 1964, and published by the Government Information Office. Originally known as the *Free China Journal*, then later changed to the *Taipei Journal*. Since 2003 issued under its present name, and in hard copy, mail edition, and an online edition.

Taiwan News (www.etaiwannews.com): A daily online newspaper (also published in hard copy form in Taiwan), one of the oldest English-language newspapers in Taiwan (formerly called the *China News*), but with perhaps the most useful website of any of the English papers in Taiwan.

Taiwan Review. A monthly glossy newsmagazine, published by Kwang Hwa Publishing Co, with a branch in Los Angeles. Formerly known as the *Free China Review*. Lavishly illustrated with excellent full-colour photos, and sometimes maps and graphs, the articles cover the spectrum, with each issue having a special focus in addition to other articles.

Index

2–28 Incident 184n28

Aborigines 10, *11*, 13–15, 26, 136–57, 183n12, 188n81, 192n38; Ami 128, 200n33; Atayal 125, 136, 138, 143, 154, 156; Bunun 124, 156; Paiwan 156; Pingpu 192n38; Rukai 156; Saisiyat 125, 154; Sakizaya 200n33; Truku 154, 156; Yami 72
Ali Shan 16, 61, 128, 183n23
Amorgos 96, 109, 193n50
Association for Relations Across the Taiwan Strait (ARATS) 29, 185n44
Atomic Energy Council Taiwan (AEC) 73, 191n11

Binnan Industrial Park 104
Bureau of Water and Soil Conservation 93, 145

Challenge 2008 53, 173–4
Chang, Juu-en 98, 102
Chang, Kuo-lung 78, 98
Chang, Lung-shen 96
Changhua 46, 76, 115, 131, 162, 163, 182n6
Chao, Shao-kan 85, 96
Chen, Shui-bian 1, 78, 84, 85, 96, 97, 105, 118, 119, 154, 170, 173, 176, 177, 191n3, 191n12, 192n31, 192n38, 201n10, 201n16, 202n17
Chernobyl 28, 33, 71, 191n10
Chiang, Ching-kuo 28, 84, 137, 140
Chiang, Kai-shek 17, 22, 28, 137, 199n10
Chiayi *2*, 16, *24*, 61, 64, 67, 93, 162, 163, 182n6, 198n4
Chien, Yu-hsin 95–6, 102
China 3, 7, 9, 10, 20–2, 23, 28–9, 38, 48, 49, 50, 53, 58, 60, 62, 66, 68, 70–1, 84, 85, 86, 89, 117, 125, 126, 127, 132, 133, 136, 158, 166, 173, 176, 178, 179, 184n30, 184n34, 185n44, 187n54, 189n89, 190n133
Commission for Conservation of Southern Bluefin Tuna 66
Construction and Planning Administration 93, 95, 96, 126
Council of Agriculture 58, 93, 127, 129, 152, 153, 159–61, 163
Council of National Development and Science and Technology 94
Council for Aboriginal Affairs *see* Council for Indigenous Peoples
Council for Economic Planning and Development 96, 153
Council for Hakka Affairs 94
Council for Indigenous Peoples 94, 153, 192n38, 200n34
Cross-Island Highway 75, 121, 125, 135–57

Democratic Progressive Party (DPP) 1, 28, 75, 78, 83, 84, 85, 86–7, 89, 98, 118, 119, 150, 151, 170, 173, 191n3, 192n22, 192n31, 193n44
Department of National Parks 96
Dongsha Marine National Park 126–7
Dutch 12–13

Energy Bureau 130
Environmental Impact Assessment Commission 104–5, 194n64
Environmental Performance Index (EPI) 31
Environmental Protection Bureau 84, 90, 93
Environmental Quality Protection Foundation 30, 47, 109, 134

Environmental Sustainability Index (ESI) 30
Erjen River *5*, 46
Europe 31, 47, 50, 71, 88, 110, 116–17, 120; Eastern Europe 71, 74
Executive Yuan 94, 133, 151, 153, 154, 155, 178, 192n41

Feitsui Reservoir *17*, 60, 109, 194n52
Fengshan Reservoir 109
Forestry Bureau 61, 62, 93, 127, 129
Formosa Plastics Group (FPG) 48, 77, 133, 178–9
Fujian 9, 21, 125, 182n4, 190n133
Fushoushan 140, 156, 199n10, 200n40

Germany 50, 110
Green Island (Lu Tao) 128, 182n4
Green Party Taiwan 86, 88–9
Green Silicon Island 1, 87, 88, 170, 172, 180
Guangdong 9, 21

Hakka 9–10, 13, 89, 96, 156
Han 13, 154
Hanyu pinyin 183n11
Hau, Lung-bin 85, 97, 193n52
Hoklo 9–10, 13, 89, 96, 98, 156
Hong Kong 20–2, 28, 38, 55, 174, 179–80, 184n30, 185n12, 185n13, 188n74
Hsinchu *2, 24*, 93, 113, 115, 126, 148, 182n6; Science-Based Industrial Park 20, *27*, 113–14, 177
Hsintien River 39, 42
Hsuehpa (National Park) *see* Sheipa
Hsuehshan Tunnel 196n112
Hualien *2*, 15, *24*, 43, 47, 63, 89, 118–19, 121, 125, 128, 137, 161, 183n21, 196n112, 200n33, 200n40
Hushan Dam 119, 163

Ilan *2*, 15, *24*, 77, 93, 97, 126, 138, 148, 196n112, 200n40
International Atomic Energy Agency (IAEA) 73
International Commission for the Conservation of Atlantic Tunas (ICCAT) 66

Japan 7, 9, 12, 14–16, 19, 20, 25, 29, 31, 33, 38, 45, 47, 50, 55, 57, 60–61, 65, 66, 73, 74, 83, 99, 104, 110, 118, 120, 122, 128, 136, 141, 156, 158, 160, 174, 179, 183n18, 183n20, 183n21, 183n23, 188n74, 189n89, 198n4
Joint Commission on Rural Reconstruction 19, 192n43

Kaohsiung *2*, 4, 15, 20, 23, *24*, 25, 35, 38, 43, 45, 46, 48, 53, 54, 55, 64, 76, 77, 89, 93, 108, 109, 110, 111, 119, 120, 126, 127, 156, 166, 177; Love River 42, 108
Kaoping River *5*, 119, 169
Keelung *2*, 15, 23, *24*, 93, 114, 108; river 39, 42, 59, 60, 127
Kenting National Park 96, 124, 128
Kinmen National Park 125–6, 182n4
Kuandu Nature Reserve 127
Kukuan 147, 148
Kuokuang Project 178
Kuomintang (KMT) 17, 21, 22, 23, 25, 26, 28, 61, 71–3, 76, 77, 78, 80, 83–4, 86, 87, 89, 97, 100, 118, 122, 125, 140, 143, 153, 184n28, 184n34, 191n12, 192n22, 192n23, 192n26
Kyoto Accord (Protocol) 87, 131–3, 198n155

Lanyu Island 72, 131–2, 175, 182n4
Lee, Teng-hui 28, 84, 85, 87, 96, 97, 177, 192n22
Lee, Yuan-tseh 175
Legislative Yuan 85, 96, 97, 100, 192n41
Lien, Chan 85, 86
Lin, Jun-yi (Edgar) 78, 96
Lishan 138, 140, 142, 144, 148, 199n7, 200n42
Lukang 76; rebellion 76

Mainlanders 10, 15, 17, 23, 26, 89, 96, 97, 137, 145, 156, 182n4, 184n28, 188n81
Makao National Park 126
Malaysia 48, 127
Mandarin 10, 22
Matsu 49, 73, 182n4, 190n133
Meinung Dam 120
Miaoli *2, 24*, 109, 115
Minamata 201n5
Mingde Reservoir 109
Ministry of Economic Affairs 93, 105, 129, 131, 163, 178; Energy Commission 93; Industrial Development Bureau 93, 105; Water Conservancy Agency 93, 163; Water Resources Bureau 93
Ministry of Education 93
Ministry of Environment and Resources 94, 153
Ministry of Interior 83, 93, 129, 152, 153
Ministry of Ocean Affairs 95

Ministry of Transportation and Communications 129
Montreal Protocol (1987) 70
Mountain Agricultural Resources Development Bureau (MARDB) 143, 144, 145

Nantou 63, 148, 156, 182n6
National Council for Sustainable Development 31, 91–2, 94, 163, 174, 180–1, 201n10, 202n20; Sustainable Development Index 180–1
National Land Recovery Plan 150–7
National Park Service 84
New Party 83, 85–6, 97
Newly Industrialized Economies 20, 184n30
Non-Governmental Organizations (NGOs) 75, 76, 78–83

Pan-Blue Alliance 83–6, 89, 132, 151, 155, 185n5, 192n26, 192n38, 200n34
Pan-Green Alliance 85, 86–9, 132, 151, 155, 185n5, 200n34
Penghu Islands 128, 131
People First Party 83, 85, 89
Peoples Republic of China (PRC) *see* China
Philippines 7, 12, 21, 124, 127
Pingtung *24*, 35, 110, 119, 127

Qing Dynasty 13–14, 60, 83, 124, 126, 138, 183n18

Republic of China (ROC) 10, 17, 21, 23, 26, 60–1, 65, 86, 90–1, 97, 125, 135, 136, 137, 144, 154, 160, 182n4, 183n20, 184n34, 190n133, 193n44, 198n4

Sheipa National Park 61, 64, 120, 125, 126, 200n40
Shihmen Reservoir *17*, 109, 148, 156, 162, 164, 169
Singapore 20, 179–80, 184n35, 188n74
Soong, James 85
South Korea 3, 12, 20, 31, 38, 65, 73, 158, 179–80, 184n35, 189n89
Straits Exchange Foundation (SEF) 28, 185n44
Suao 118–19, 137
Sun Moon Lake 15, 109, 128, 196n119

Tachia River *5*, 26, 64, 125, 136–57, 200n40
Taichung *2*, 15, *24*, 25, 43, 47, 77, 93, 109, 111, 131, 136, 143, 177, 182n6, 193n44, 198n4
Tainan *2*, 16, *24*, 45, 46, 64, 89, 93, 105, 109, 119, 121, 131, 138, 177, 191n3, 198n4
Taipei *2*, 4, 15, 17, 23, *24*, 25, 36, 38, 39, 40, 42, 47, 50, 53, 54, 55, 59–60, 63, 64, 74, 75, 76, 77, 78, 80, 82, 89, 93, 97, 106, 107, 108, 109, 110, 111, 114, 120, 124, 126, 127, 166, 192n38, 193n44, 196n112, 201n16; Hsichih (Sijhih) 59, 108; Rueifang 108; Sanchung 60; Wanhua 189n97; Zoo 63
Taitung *2*, 15, *24*, 43, 62, 89, 128, 183n21
Taiwan Environmental Protection Administration (TEPA) 29, 36, 37, 39, 40, 42, 44, 45, 46, 47–53, 78, 81, 90–1, *92*, 95–105, 108, 129, 133, 165, 169, 174, 179, 185n45, 190n2, 194n55, 198n159, 198n160, 202n24
Taiwan Environmental Protection Union (TEPU) 78, 80–3, 88, 98, 103, 117, 176, 191n7, 191n18
Taiwan Power Company 78, 131
Taiwan Provincial Government 93, 193n44
Taiwan Solidarity Union 83, 87, 89, 176, 177
Taiwan Sugar Corporation 17, 156
Tanshui River *5*, 39, 42, 60, 75, 105, 108, 127
Taoyuan *2, 24*, 44, 64, 93, 109, 115, 119, 126, 138, 148, 162, 164, 177
Taroko National Park 61, 75, 120, 125, 126, 142, 154
Tayuling 138
Techi (Deji) Dam 143, 144, 147
Ten Nightmares, the 87, 117–21, 196n112
TEPA *see* Taiwan Environmental Protection Administration
TEPU *see* Taiwan Environmental Protection Union
Three Mile Island 28, 71, 191n10
Tongyong pinyin 183n11
Tsai, Hsung-hsiung 96
Tsengwen Reservoir 120

United States 20, 25, 31, 38, 40, 47, 48, 50, 58, 64, 68, 71, 73, 74, 88, 99, 104, 110, 116, 118, 122, 160, 168, 172, 175, 179, 183n22, 186n28, 189n89, 192n41, 192n43, 193n44, 193n49, 199n11; Love Canal 33

VACRS *see* Vocational Assistance Commission for Retired Servicemen
Vietnam 21, 127
Vocational Assistance Commission for Retired Servicemen 26, 137, 140, 142, 143, 154, 156–7, 199n10, 200n42

World Economic Forum 179–80; Global Competitiveness Index 179–80
World Trade Organization (WTO) 57, 108, 155, 157–62
Wuling 140, 156, 200n40
Wushantou Reservoir 198n4
Wushe 138, 148, 156

Xiamen 21

Yangmingshan National Park 120, 124–5, 196n119
Yen, Chia-kan 184n41
Yu Shan 3, 63, 75; National Park 61, 120, 124, 125
Yunlin 44, 67, 77, 119, 162, 163, 178, 182n6